THE COMPLETE GUIDE TO
DRONES

BUILD + CHOOSE + FLY + PHOTOGRAPH

THE COMPLETE GUIDE TO
DRONES

WHATEVER
YOUR
BUDGET

ADAM JUNIPER

ilex

COMPLETE GUIDE TO DRONES

An Hachette UK Company
www.hachette.co.uk

First published in the United Kingdom in 2015 by
ILEX, a division of Octopus Publishing Group Ltd

Octopus Publishing Group
Carmelite House
50 Victoria Embankment
London, EC4Y 0DZ
www.octopusbooks.co.uk

Publisher: Roly Allen
Associate Publisher: Adam Juniper
Senior Specialist Editor: Frank Gallaugher
Senior Project Editor: Natalia Price-Cabrera
Assistant Editor: Rachel Silverlight
Art Director: Julie Weir
Designer: Jon Allan
Senior Production Manager: Peter Hunt

ISBN 978-1-78157-307-5

A CIP catalogue record for this book is available
from the British Library

Printed and bound in China

10 9 8 7 6 5 4 3 2 1

CONTENTS

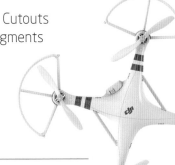

INTRODUCTION

Call them what you will (and long after the ship has sailed that is still a matter of debate for many), Drones have caught the public imagination. We've all heard how these buzzing wonders will shoot through the skies delivering pizzas and packages, enabling amazing feats of photographic wonder and many more applications beside.

We've also seen the word applied to unmanned military craft dealing death with debatable discrimination, but that is a matter for other pages and, no doubt, many years of hearings. On the other hand, believe it or not a military drone program was instrumental in launching Marilyn Monroe's career!

But what can we actually do with drones? How do you choose the right one? How will it work? How do you maintain it? What are the rules? This book is set to answer all these questions and more, or at least show you exactly where to look to find exactly the right answer for yourself.

MARILYN MONROE

Norma Jean Dougherty (aka Marilyn Monroe) working at the Radioplane Munitions Factory in Van Nuys, CA, building target-practice drones in 1944. Photographed by David Conover for the military and seen by Ronald Reagan, Conover's C.O, this photo helped kick-start Monroe's career.

INTRODUCTION

This book is not a manifesto for self-building or another well-rehearsed media event for one of the major players providing off-the-shelf copters. Personally I think both have their merits (and I know people who can only imagine one or other).

If you are interested in building, though, Chapter 5 includes step-by-step instructions for constructing your own drone; we'll show you exactly where you can get all the parts you need to get in the air as cheaply as possible, and in the spirit of community, we strongly urge you to share your results with Jack (its designer) and I with the hashtag #tamesky1 (it'll make more sense when you've read it).

DJI PHANTOM
This Phantom 3 continues the incredibly popular series from DJI. The shell design is almost the first thing many bring to mind when they think of drones.

And on that note, let me be the first to welcome you to what is an amazing, limitless community.

Have fun, fly safely, and respect others!

CHAPTER 01
DRONE
BASICS

**Where drones came from,
and what's up there now!**

HISTORY OF DRONES

Multicopters, or drones, were something of a minority interest until a little known car entertainment company, Parrot, grabbed the world's attention with the AR.Drone at the 2010 International Consumer Electronics Show (CES) in Las Vegas.

PARROT AR.DRONE
Shown with the now-iconic indoor safety hull fitted.

PARROT'S 22-INCH (57CM) multicopter was a true consumer product that was ready to fly, right out of the box. It was constructed from replaceable nylon and carbon-fiber parts, with interchangeable hulls made of lightweight polystyrene. It also housed a small, front-mounted camera.

However, it was arguably the fact that the 'copter was controlled from an iPhone, using Wi-Fi, that meant it captivated a whole new crowd. Journalists at CES were only just getting used to a world with the iPhone App store (which was itself only 18 moths old at that point), and this remote-controlled device appealed across the board. With it's app control it attracted the attention of Apple, iPhone, and mobile telephony journalists, while its game-like control interface meant that reporters from the video games industry saw it as their domain—something Parrot encouraged by offering games with virtual objects the pilot could see only on their phone's screen. Meanwhile, photography writers were drawn to the "flying camera" angle (even though the camera's resolution was relatively low), and futurologists were drawn by the "augmented reality" of the virtual objects game.

In short, the AR.Drone got a lot of press. It probably didn't hurt that the signature indoor hull with black polystyrene guards around each prop not only looked great flying around an exhibition center, but was also slightly reminiscent of the Samson gunships seen in the previous summer's smash hit movie Avatar.

With somewhat less fanfare, developers who were already working on an open source autopilot system for fixed-wing model aircraft began work on ArduCopter in the summer of 2010. This was released the following summer and enabled enthusiasts to construct their own quadcopter, similar to the way in which they could already build other model aircraft.

The "Ardu" part of the name derived from Arduino, the name of the company that appeared in 2005. The company's aim was to supply small computer systems designed to help hobbyists, students, and professionals build inexpensive devices that could interact with their environment. As Arduino boards are open source, they lend themselves perfectly to being programmed for new applications—coding itself is beyond the scope of this book, but we'll look at the starting points later on.

Chris Anderson, then editor of Wired magazine, built on Arduino in 2007 to create the ArduPilot (and now ArduCopter). He created a community site, DIYDrones.com, for support the project. DIYDrones.com has since added a commercial arm—3DRobotics—which sells ready-build 'copters and fixed-wing aircraft, as well as its own flight controllers.

3DRobotics does not have the market to itself, though. By 2013, Chinese company DJI had built up an impressive array of components for enthusiasts and professionals, including one of the most respected

NINTENDO WII

The 2006 success of the Nintendo Wii with it's motion-sensing controller inspired many enthusiasts to create projects that could use those sensors. One such was MultiWii, which operates on an Arduino board and the board extracted from a Wii controller (which is a cheap source of motion-sensing components).

ARDUINO PRO

Costing the same as about three franchise coffee bar grandes, the Arduino Pro is able to communicate with a variety of digital and analog sensors. Properly programed, it can receive signals from sensors and send them to motor controllers, making it perfect for taking charge of a multicopter.

flight controllers on the market. However, the launch of its Phantom quadcopter saw it go on to capture a massive slice of the 'copter market as well.

To grow this market, DJI realized that it needed to appeal to a group who would find real needs for the company's 'copters, not just enthusiasts. That group was creatives looking for an aerial photography and video solution, so DJI set about making the Phantom easier to use, with improved picture-taking capabilities.

In addition to the product itself, DJI's marketing blitz was bigger than anything that came before it, with the Phantom's "halo effect" (and people's suspicion of it) still being felt across the industry.

DJI PHANTOM ADVANCED

When many people hear the word "drone," the first thing that springs to mind is the devastating use (and misuse) of powerful military aircraft. The best known of these is the General Atomics MQ-1 Predator, which entered service in 1995, but this was not the first time the military took an interest in unmanned aircraft.

THE EARLIEST RECORDED account of unmanned military craft being used dates all the way back to August 1849, when Austrian forces were laying siege to Venice. The Austrians loaded balloons with bombs, and when the wind carried them over Venice, they released them using an electromagnet on the end of a very long copper wire.

This is a far cry from today's use of unmanned aircraft, which can be traced back more directly to WWI. In 1917, Archibald Low, an English engineer and head of the Royal Flying Corps Experimental Works, built a radio controlled plane known as the "Ruston Proctor AT" (Aerial Target). The intention was to place explosives on the plane and fly it into a target, making this the first cruise missile.

At the end of WWI, Reginald Denny—another Englishman who had served in the Royal Flying Corps—became interested in radio controlled model planes. This led to him setting up a model plane shop in 1934 and developing his own line of "Dennyplanes." The company subsequently won a contract with the US Army, which saw them manufacture and supply almost 15,000 Radioplane target drones (as the remote controlled planes were known), which were used for training anti-aircraft gunners.

Unmanned aircraft also featured during the Vietnam War, when the 4080th Strategic Recon wing of the USAF launched drones from adapted C-130 transport planes. The drones would fly over the designated

PREDATOR

Including the updated "gray eagle" model, this drone has been in service since 1995 with the USAF. Although it was initially a reconnaissance aircraft, from 2004 the MQ-1A "Predator" has carried Hellfire missiles. It's an effective weapon, but not without its problems; many have been lost and, including its successor—the MQ-9 "Reaper"—four have been shot down.

target area taking photographs, then parachute to the ground once they were in safe territory, ready to be picked up by helicopter.

However, it was Israel's 1982 invasion of Syria that demonstrated more offensive uses for unmanned aerial vehicles (UAVs). As well as flying reconnaissance missions, Israel used drones as electronic decoys and jamming tools, helping them to secure a firm victory over the Syrian Air Force.

After that, it was inevitable that other military forces would take unmanned aircraft more seriously, leading to the arrival of heavy craft, such as the American Air National Guard's fleet of more than 40 Predator drones, the 100+ Reapers and Global Hawks in the hands of the USAF, and the UK's stealth-drone Taranis. All of these UAVs are capable of semiautonomous flight, with a payload of missiles that can be fired via remote control.

The rapid development of this type of aircraft has put a great deal of strength into the hands of a select few governments. However, while command and control is typically straightforward (via encrypted remote control), the legitimacy of targeting seems to be open to some serious doubts.

2002 saw the first fatal US missile attack, which resulted in the death of a civilian scrap-metal merchant who was selected as a target by the CIA at least in part because he was "about the same height as Bin Laden."

You can form your own judgment on the morality of this (and more than 2,000 other deaths caused by the US drone program), but it's worth noting that the Pentagon's spokesperson, Victoria Clarke, said of the attack: "We're convinced that it was an appropriate target." She then went on to add: "We do not know yet exactly who it was." With this level of exactitude, it is perhaps hardly surprising that the word "drone" is not always one that people feel positive about.

NASA ALTAIR
A NASA-adapted variant of the Reaper, used by NASA's Earth Science Enterprise. With the cover removed, its satellite antenna is revealed.

PREDATOR PILOTS
Captain Richard Koll and Airman 1st Class Mike Eulo piloting a MQ-1 Predator Drone over Iraq. Once the drone has taken off and its flight checks been performed by this team, control is handed over to a base in the US.

BAE TARANIS
Developed to British government specifications, the Taranis, which first flew in 2013, is fast and stealthy, with shielded weapons bays to reduce radar detection. It is intended to be able to "think for itself" and defend itself against manned and unmanned attacks.

WHO'S FLYING MILITARY DRONES

At time of writing, countries known to fly military drones include:

United States	Russia
United Kingdom	India
Israel	Iran
France	Turkey
Germany	UAE
Italy	Morocco
China	Spain

THE EMERGENCE OF MULTICOPTERS

Just as the idea of UAVs has a longer lineage than you may have imagined, so multicopters have a significant heritage, with some of the earliest successful vertical takeoff and landing aircraft (other than balloons and zeppelins) including manned quadrotors.

AS EARLY AS 1907—more than 30 years before the first helicopter—French brothers Louis and Jacques Breguet flew what was essentially a giant quadcopter. Their Gyroplane No. 1 was piloted by Monsieur Volumard, although "piloted" is perhaps a generous term: although it was fully able to support its own weight, the craft had to be controlled with ropes from the ground.

The design had four blades on each rotor, with each blade measuring 8 meters in length. The rotors were then doubled up in the manner of a biplane, to give a total of 32 blades. Gyroplane No. 1 was created in collaboration with Professor Charles Richet, a true inter-disciplinarian who would go on to be awarded the Nobel Prize for Physiology for his work on anaphylaxis, and who also wrote much on parapsychology—including *Our Sixth Sense*—and coined the term "ectoplasm." Certainly a sixth sense would have been required to fly the Breguet-Richet gyroplane untethered.

By 1921, another Frenchman, Étienne Oehmichen, added a fifth control propeller to a similar concept and by 1923 was able to fly 575 yards (525m) in a surprisingly stable fashion.

Oehmichen's great rival at the time was the Argentine inventor, Raúl Pateras Pescara, who bested him soon after with a design that looked more like a traditional helicopter, with variable pitch rotors above the pilot.

In the same year, George de Bothezat, a Russian emigree who fled to America in 1918, built a quadrotor for the US Army Air Service at Wright Field. Although successfully flown over 100 times, the project was ultimately canceled, as the so-called "flying octopus" was directed more by the wind than any effort on the pilot's part.

Back in France, Oehmichen had picked up a 90,000 Franc prize by flying his craft in a defined triangle, returning to the point of take off over 7 minutes later. However, it was Pescara's approach that was to dominate explorations into vertical takeoff and landing (VTOL).

While Pescara's "helicopter" would form the rough blueprint for modern helicopter designs, notable VTOL exceptions appear in military experimental aircraft, including the British Cierva W.11 Air Horse from 1948, the Curtis-Wright VZ-7 "Flying Jeep" (the US Army had two prototypes in service from 1958 to 1960), and the Piasecki VZ-8 Airgeep that it competed with.

GEORGE DE BOTHEZAT
In 1923, De Bothezat's multicopter managed to stay aloft for 2 minutes 45 seconds and land safely.

CIERVA W.11 AIR HORSE

When it began testing in 1948, the Air Horse was the largest ever rotary-wing aircraft. All three of its rotors span in the same direction; the somewhat low-tech solution to avoiding spinning was to angle the motors in the opposite direction.

CURTISS-WRIGHT X-19 / X-200

A single engine propelled all four rotors in the X-100 prototype. The engine was directly linked to the front two rotors, with the exhaust gases turning the rear blades when they were pointed upward. The X-19s added an engine, but after one of the two prototypes crashed in 1965, the program was canceled.

X-22

The four ducts/nacelles around the X-22's tiltable rotors gave it a more than unusual appearance; each rotor had its own gas turbine engine mounted on the rear wing. Even though it didn't meet the military's target speed of 326mph (525km/h), Cornell Aeronautical Laboratory kept it in flight until 1988.

MV-22 OSPREY

First tested in 1989, the V-22 entered service with the US Marines in 2007 and with the Air Force in 2009. Like the X-19 and X-22 before it, the MC-22 has tilt rotors allowing VTOL and high-speed flight. It can even refuel in mid-air when flying with the rotors tilted forward.

However, one of the most significant developments came with the X-22 in 1966, which featured an on-board computer designed for "stability augmentation" (which was achieved by altering the blade pitch). Although the first X-22 crashed, the pilots were fine and vibration issues were resolved with the second iteration of the craft, which went on to clock up over one hundred hours in the air. This testing period meant designers were able to program the on-board computer with numerous algorithms, effectively providing them with a simulator for other aircraft, both real and theoretical.

At the time, this computer concept cost the military over $40 million, but this idea of taking the complex task of controlling individual motors out of the pilots hands is what makes modern multicopters possible. Indeed, the small size of radio controlled quadcopters—and their consequent lower

stability—means that an electronic flight controller is the only practical way of flying them.

The essential component to these flight control systems is a gyroscope, a technology that has been improving in parallel to the craft that depend upon it. Already small enough for aircraft systems by WWII, they were miniaturized still further for the guided missile era, until they measured around 1 inch (2.5cm) in diameter and weighed in at close to 3 oz (85g). Since the early 1990s, all-digital MEME gyroscopes can be added to a circuit board—this is exactly the same technology that tells your smartphone which way up it is being held. These lightweight gyros are what allowed the first RC helicopters to be controlled by the "average" pilot, using the tail rotor to control any spin automatically. And now, with these gyros in the toolkit, all that was needed was a multicopter to put them in.

RC LEGACY

Multicopters are radio controlled (RC) aircraft. However, it's not just their otherworldly hovering capabilities that makes them different to the models that preceded them—the new 'copters are a product of the digital era and can address the problems of flight in new ways.

THE EARLY DAYS of RC flight were dominated by gunnery target drones used for military training purposes. However, in the late 1930s a steady stream of hobby aircraft began to appear as enthusiastic amateur (or "ham") radio operators began to look for other uses for their skills with vacuum-tube radio technology.

In 1937 the burgeoning industry went public with a US national competition, although of the six entrants only three flew, and only one — Chester Lanzo—flew for longer than ten seconds. A year later, Walt and Bill Good flew to first place with their balsa wood plane, "Guff," which was so successful that the original has lived in the National Air and Space Museum in Washington, D.C., since 1960.

Those early RC pioneers needed to be expert craftsmen and electronics whizzes. Even then, their bulky, hand-made controls were often mounted on tripods because of their weight, and often offered nothing more than on/off control on a single axis (yaw, via the rudder).

The widespread arrival of the transistor in the 1960s reduced the size and weight of RC systems and a commercial market started to emerge. As it grew, controllers that worked on multiple channels became more common, with each additional channel allowing the a different device on the model to be controlled remotely; adding ailerons to control the plane's roll, for example, as well as the rudder.

Governments have long reserved the use of specific radio bandwidths for different uses, and this makes it easier for RC controller manufacturers to work to standards. More usefully, it is also why your mobile phone works and why your car radio doesn't pick up military communications.

The older bands used a system of matching crystals, so you would have, for example, a channel 12 (or other number) crystal for both your controller and your aircraft. The downside to this system was that if anyone in your area happened to be using the same channel your signal would be jammed and you would crash.

This changed around the turn of the millennium, with the arrival of the 2.4GHz "spread spectrum," which allowed automatic sharing of the same channel by multiple users.

RC PLANE

Flying a traditional RC plane requires the same basic skills as piloting a light aircraft, although it has the advantage of not being inside if something goes wrong! Nevertheless, unless you master landing early on, it can be a very expensive hobby.

At the same time, the control devices on aircraft have been improved and refined, with blunt, 2-position reed switches usurped by servo motors that can be more precisely controlled for delicate turns and maneuvers. This has also been joined by a standard configuration of four channels to control the rudder, ailerons, elevator, and throttle.

However, despite these advances, the basics of flight remain the same: when pulled through the air by the propeller the aircraft's aerodynamic shape provides lift and the craft is steered using its control surfaces.

That system is the legacy that the RC world bequeathed multicopters, and hobby shops are filled with radio control systems designed with RC aircraft (as well as cars and boats) in mind.

AIRCRAFT CONTROL SURFACES

Traditional airplanes rely on carefully positioned control surfaces: ailerons at the wings' tips (which swing in opposing directions to roll the plane) and a rudder to turn it (apply yaw). Lift comes from the shape of the wing, and the amount of lift is affected by the thrust from the motor. Elevators in the tail can be used to push the nose up or down (they swing in the same direction as each other).

When you're in a commercial airliner, you will see that the pilot extends flaps for takeoff and landing (where higher lift at a lower speed is required) and then retracts them at cruising speed for aerodynamic smoothness. These are less common in RC planes.

HOW MULTICOPTERS FLY

The lift that keeps a traditional plane aloft comes from the air flowing over its carefully crafted shape, meaning forward momentum is essential. Multicopters are a lot more flexible—only the propeller blades need to be aerodynamically shaped, with control achieved through subtle changes to the speed of their rotation.

THE CURVED SHAPE of airplane's wing smashes into the air as it hits it. The upward curve means the shape is more disruptive of the air above the wing, pushing it away and effectively lowering the air pressure above the wing in comparison to that below where the shape remains flat and air is unimpeded. Once the difference in pressure above and below the wing is enough, the higher pressure below has the effect of pushing the wing upward. Of course, this only works when the wing is moving forward fast enough for it to create the necessary lift, which is why long runways are needed for take off.

With helicopters, a similar curve is applied to blades rotating above the cockpit, which is why they are often called "rotary wing aircraft." The angle of helicopter blades can be adjusted, and the tail rotor can be sped up or slowed to turn the cockpit, a little like the rudder on a plane. However, the tail rotor has another function; without it there would be nothing to stop the cockpit spinning out of control with the blades.

Most multicopters eschew such complicated mechanics in favor of making very fast alterations to the rotation speed of an even number of matching propellers. Common prop arrangements are shown below, but they're far from the only options. The biggest advantage of having an even number of propellers is that their opposing rotations very easily eliminate the natural effect of torque rotation that would otherwise cause the craft to spin (this is the same reason Leonard da Vinci's helicopter would never have flown, even with a motor replacing the "four strong men" he suggested).

Rather than achieving control through complicated surfaces, the multicopter simply adjusts the speed of its props to lean into the pilot's chosen direction or to temporarily take advantage of the torque effect to make turns. By taking tilt readings from on-board gyroscopes (often called "6-axis gyro") and directional readings from a compass, the computer can ensure that the right amount of thrust is applied by each rotor.

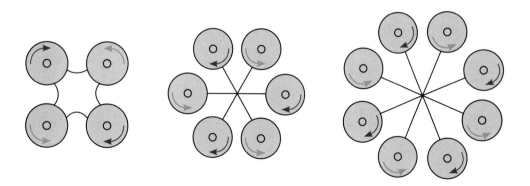

OPPOSING ROTORS

In general, featuring an even number of rotors that rotate in opposing directions avoids the torque effect that would otherwise cause a multicopter to spin out of control. However, there are some exceptions to this rule, as shown on page 33.

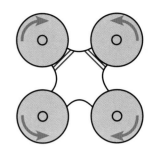

HOVER

A stable hover is achieved when the craft's thrust is being directed downward. All things being equal, the props will be rotating at the same speed.

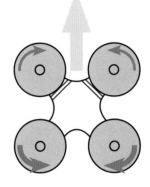

DIRECTIONAL CONTROL

Were the computer in charge, there would be no real reason for a 'copter to have a designated front, but for more natural piloting (and often because of camera positioning) multicopters will have a clearly marked forward direction. The DJI Phantom pictures here uses red marking stripes and colored LED lights.

DIRECTIONAL FLIGHT

Directional flight is achieved by pushing the craft forward. To begin movement, the copter will lean into the direction of travel and hold this angle to maintain forward momentum.

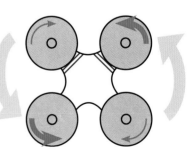

ROTATION

A multicopter can rotate around its central axis while hovering by increasing the speed of motors in the appropriate direction. From a user's perspective you simply need to move the yaw control left or right.

THE BASICS OF MULTICOPTER FLIGHT

Despite a lot of ominous stories in the less-well-researched press, multicopters are generally flown using similar principles to model airplanes, which have been inoffensively buzzing around the skies for years. All you need to do is master the basics of radio control.

PILOT'S CONTROLLER

Some drones come with specially-made remote controls, but most use off-the shelf RC controllers that use publicly available radio frequencies to give you up to 2 miles range. These controllers feature sticks, switches, and dials that can be configured for remote controlled planes, cars, and boats, as well as drones. That has advantages and disadvantages, as you'll see on page 46.

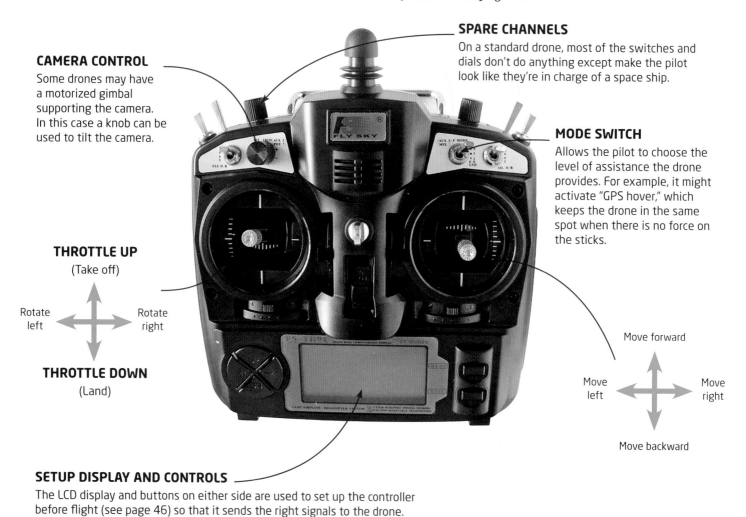

CAMERA CONTROL
Some drones may have a motorized gimbal supporting the camera. In this case a knob can be used to tilt the camera.

SPARE CHANNELS
On a standard drone, most of the switches and dials don't do anything except make the pilot look like they're in charge of a space ship.

MODE SWITCH
Allows the pilot to choose the level of assistance the drone provides. For example, it might activate "GPS hover," which keeps the drone in the same spot when there is no force on the sticks.

THROTTLE UP
(Take off)

Rotate left — Rotate right

THROTTLE DOWN
(Land)

Move forward

Move left — Move right

Move backward

SETUP DISPLAY AND CONTROLS
The LCD display and buttons on either side are used to set up the controller before flight (see page 46) so that it sends the right signals to the drone.

THE MULTICOPTER DRONE

With its arrangement of clockwise and counterclockwise spinning props, a multicopter can be built in an endless variety of ways. They can be bought boxed and "ready to fly," as "nearly ready to fly" kits that have all the right parts but require some assembly, or they can be lovingly assembled and upgraded with parts from your local hobby store.

TRANSMITTER (OPTIONAL)
Transmits a live signal from the camera to a ground station.

CONTROL BOARD
The drone's "brain." The control board is connected to a compass, accelerometer, and GPS sensor, and translates signals to make the drone fly.

ESC
The Electronic Speed Controller sends power to the motors according to the control board's instructions.

RECEIVER
Receives radio controls.

BRUSHLESS MOTORS
Their speed changes constantly to keep the craft upright.

PROP
Props can be made in many colors and materials, including carbon fiber and wood.

CAMERA (OPTIONAL)
Many drone users choose to mount a camera, but there is a weight penalty. This is a GoPro camera, which can record or link to a transmitter for live video.

FOOLS RUSH IN (AND A FOOL AND HIS DRONE ARE SOON PARTED)

The best safety and money saving tip is to practice with a low-cost drone before putting an all-singing, all-dancing hexacopter with a GoPro camera on your credit card. Hubsan sell tiny quadcopters and controllers that cost less than a video game, but they react in exactly the same way to the control sticks as more advanced 'copters. By the time your mega-drone lifts off you'll be a pro!

The Hubsan x4 could land on your palm (if you were good enough)

INDOOR AND "TOY" DRONES

Multicopters are a lot of fun, and the best, cheapest, and safest place for every member of the family to experience them—especially for the first time—is in the home.

ACCESSIBLY PRICED DRONES have been made possible because their stability system shares so much in common with mobile phones. The same gyro that makes the screen and camera switch orientation when you turn your phone is used to keep your drone stable, and the same magnetometer (compass) that your phone uses to determine the direction it is pointed in is equally useful to some drones (although others rely on the pilot). The development of tiny, powerful, and fast-charging batteries also owes a lot to mobile phone technology, although the cells found in these minidrones are typically even smaller.

A big advantage in designing minidrones is that the pilot will certainly be very near to the craft, so video relay is less essential and long distance RC systems can be dispensed with. Instead, manufacturers pick from a variety of established, cheap, light, and low-power alternatives, such as Wi-Fi, Bluetooth, and even Infrared control (much like your TV commander).

PARROT ROLLING SPIDER
This minidrone uses low-power Bluetooth 4 signals for control, displaying a control pad on a smartphone (right).

FUTURE "TOYS"

Although it is unlikely to be priced as a toy when on sale, the Nixie wearable drone gives an idea of the future of the technology, having won an Intel competition to develop wearable tech for Intel's new system-on-a-chip.

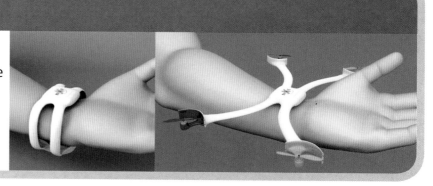

This is great news if you're starting out flying because it means you can pick something up that doesn't cost too much. More importantly, the same basic control system is found on most of these devices, so the skills you develop for avoiding obstacles in your hallway will also apply when you've invested more money in something a little more rugged and moved outdoors.

HUBSAN X4 H107D

Charging in just 30 minutes, and with a flight time of around 6 minutes, this configuration of Hubsan's popular X4 can send VGA resolution video directly to the controller and record it to a Micro SD card.

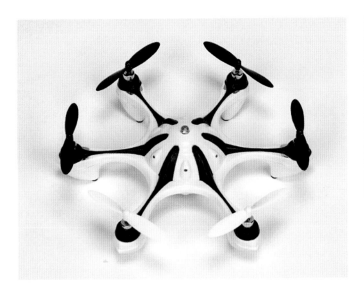

ART-TECH R11435

This hexacopter minidrone can record 720 x 480 video to a Micro SD card

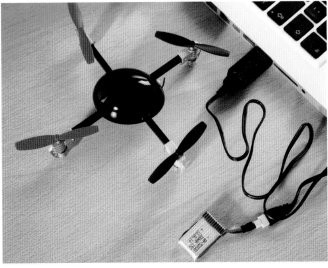

MICRODRONE

The Microdrone's battery is charged via a USB port. The included controller offers a choice of control options.

PROSUMER DRONES

01

DRONE BASICS

The drone market exploded in 2010 when Parrott—previously known for in-car entertainment systems—released the AR.Drone. Its intuitive iPhone control system, built-in camera, and accessible price stole the show at CES, arguably creating a market that has since seen more than one million drones sold.

AM I A "PROSUMER?"

Prosumers drive the consumer electronics industry; they are the people who read reviews and develop an understanding of a product and its features before they buy it, and try to get the best of it. In the world of drones, a prosumer will look for features such as a built-in camera (or the ability to mount one), but as it's still a consumer product they will still expect it to work out of the box, rather than requiring additional expense or assembly.

However, the world of the prosumer is constantly changing: whereas the AR.Drone 1.0 and 2.0 were once prosumer fare, they would now be considered more like expensive toys, especially if flown without the GPS module (an optional extra that costs around $100). This is because products such as DJI's Phantom have raised the performance standard and prosumers like "pro" quality.

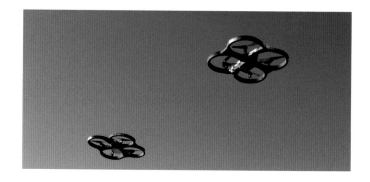

PARROT AR.DRONE 2.0

Parrot's AR.Drone 2.0, with an HD camera, probably created the prosumer market (though it's specifications make it more of a toy now). The drone features interchangeable hulls so you can fly safely indoors but keep the sail area low outdoors.

BEBOP DRONE

The BeBop (left) is the spiritual successor to the AR.Drone. Digital image stabilization negates the need for an expensive motorized camera stabilization system, and it also sticks with Wi-Fi control, rather than its competitors' RC control. However, it offers professional users a range-extender and even the ability to plug in Oculus Rift virtual reality glasses (above) so the pilot can look around the "cockpit."

DJI PHANTOM RANGE

DJI is a big name in drones, not least for its Phantom. At nearly $1000 (without a camera) this is not a cheap first drone, but it has the power to lift a professional quality camera like the GoPro and a servo-driven stabilizer gimbal (DJI's Zenmuse H3-3D gimbal is one of the best in the business and it just slots right in). DJI also offers drones with built-in, stabilized cameras of varying quality; the "Professional" model supporting stabilized 4K video.

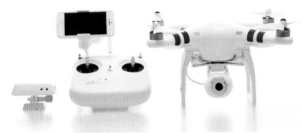

PHANTOM ACCESSORIES

Given the popularity of the product, the Phantom it is well supported with upgrades and features such as first-person video. There is also a lively third-party accessory market.

STEADIDRONE QU4D

Arguably the best thing about the drone market is the small companies who are creating beautifully built craft from off-the-shelf components, designing their own packaging and instructions. These offer prosumers a step toward self-build power with the backup of customer support.

3D ROBOTICS SOLO

Designed in San Diego, and manufactured in Tijuana, the Iris not only arrives ready to fly (including the ability to fly autonomously along paths you program), but with support for a GoPro which can also be operated by the software. 3DR have even added a port for future upgrades to the drone's bank of sensors.

BUILDING DRONES

01

For the ultimate design flexibility, the only choice is to build your own drone. By choosing your own components and attaching them to an airframe of your choice (or your own devising) you can create anything from a super-quick stunt 'copter to a stable, heavy-lift platform.

BUILDING YOUR OWN multicopter requires you to select components that will work together, connect them up, and perform any necessary setup before you are ready to fly. This sounds simple, but there is some complexity in the process—ensuring components will work together well requires a certain amount of research; the set-up can often involve quite fiddly computer software; and you'll need to master the art of soldering.

All of those subjects are covered later in this book, and in exchange for the extra effort you'll be able to make something you can be proud of. Not only that, but there's no need to spend all your money at once. You can build a basic airframe, for example, and then when you can afford it, you can add a camera and then, later, perhaps a sophisticated gimbal to stabilize the camera for top-notch video. Other additions you can make later include devices to relay flight information (telemetry) back to the ground, or perhaps even live video.

At the other end of the invisible tether, you can also make changes to your controller over time, starting out with the basic channels you need to get airborne and then add attachments to receive video and view live telemetry at a later date, or add similar receivers to your computer if you prefer.

Along with the advantages of flexibility and the ability to spread costs over time, self-building can be a really enjoyable way to pick up some useful practical skills, or a fun way to share them with a family member. When you get to fly something you built yourself, learning about electricity and developing your soldering skills is a lot more rewarding!

The downside is the uncertainty—there's a reason why so many people find Apple's "it just works" approach to computing far more appealing than purchasing a variety of components to assemble themselves! If that sounds like you, then you will likely be happier with the likes of DJI, which strives to make its consumer 'copters flight-ready straight out of the box.

DRONE IT YOURSELF
Jasper van Loenen has created a kit that can be used to convert almost anything into a multicopter.

TURNIGY MICRO-X
Not all kit drones are large—the Micro-X can rest on the palm of your hand and the circuit board forms the airframe. The kit includes motors, props, and batteries, but requires assembly.

THE SKYTANIC
The author's first hexacopter, built using an assortment of mail-order parts including a DJI NAZA flight controller and a Tarot 680 airframe. Pipe insulation around the thin carbon fiber landing struts does a lot to cushion landings.

There is a middle ground, however. Many resellers have taken the time and energy to devise packages that include all of the necessary components to build different 'copters. With one simple order you'll get everything you need. It will still be delivered as a box of boxes (and probably with different instructions from different companies), but you have the certainty that the components in question will work together: the motors and props will provide the correct amount of lift without burning out, for example. Better still, you'll often get a discount against the price of buying the components individually.

DEVELOPED KIT
A copter rigged for FPV flight with a GoPro gimbal for steady recording of those screaming race victories as well as the live FPV camera just above. LED tape on the front arms and back clearly shows the pilot where the front is and the airframe, from TBS, is built using easily-replaceable components.

PROFESSIONAL DRONES

Fully built drones that are intended for commercial use are a relatively new phenomenon. There is a certain amount of confusion over regulation, but there are already several models available, and the selection is getting better all the time.

THE AMERICAN FEDERAL Aviation Authority, or FAA, is very influential around the world, but it has been one of the slowest agencies of its kind when it comes to regulating the commercial use of UAVs (including multicopters). Luckily, equivalent agencies around the world, including the UK's Civil Aviation Authority, have been more forward looking and have created appropriate pilot certification schemes.

Not wanting to fall behind, the US Congress mandated the FAA to create regulations for commercial use of unmanned aircraft by the end of 2015. With that deadline in mind, some commercial users started to be approved toward the end of 2014.

The most obvious commercial application for multicopters is in cinematography: helicopter shots are incredibly expensive and very restrictive in terms of airspace and proximity to the cast. Filmmakers expect to budget at least $5,000 per filming day, including a pilot and camera operator. However, a professional multicopter crew working with a feature-quality camera and lens will be able to work for around $2,000 a day, possibly less. That would be for a camera operator (with a separate controller), as well as the pilot.

Compared to prosumer models, professional drones need to be able to lift much heavier cameras— high quality lenses feature a lot of heavy glass components— and stability becomes a more important consideration that maneuverability. Many professionals like to work with the added

security of 6 or more rotors which are capable of a relatively controlled landing even if they lose a motor.

Drones are useful in mapping, using the GPS feature to take photographs from specific locations which can be tiles to create a clear picture of any location, from a construction site to a farm. Aerial inspection is immensely useful for agriculture, helping farmers inspect vast areas with the accuracy required by modern high-intensity agriculture. Should the flight reveal any problems, a ground inspection can follow.

Large UAVs, such as Microdrones' MD4-3000, are targeted at public bodies such as the police (for crowd monitoring), the fire service (to reveal fire hotspots and people in danger), and commercial agencies including energy suppliers who may have hundreds of miles of cable that needs visual inspection, for example.

MD4-3000
Quadcopters can be pretty big beasts. The MD4-3000's take-off weight is 23-33lb (10.5-15kg) and without props it measures over 3 feet (1m) across. It can stay in the air for 45 minutes.

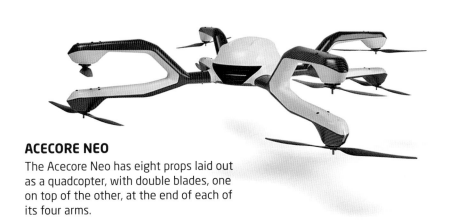

ACECORE NEO

The Acecore Neo has eight props laid out as a quadcopter, with double blades, one on top of the other, at the end of each of its four arms.

DJI INSPIRE (IN FLIGHT)

In flight, the Inspire's legs can be lifted away to allow unobstructed, 360-degree camera operation. This is common to many professional drones.

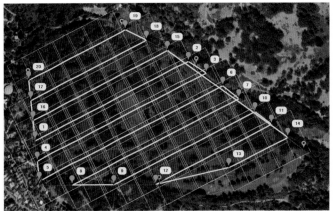

DJI INSPIRE (LANDED)

At the lower-end of the pro market is the DJI Inspire, which is capable of dual-operator use, has retracting legs, and features a 4k camera that can be swapped for an alternative payload. The Inspire also features sonar sensors to make it more responsive indoors, where there is no GPS signal to help the 'copter hold position.

MAPPING A FLIGHTPATH

Ardupilot allows you to enter some basic parameters, such as your selected camera, camera orientation, and altitude, and it will automatically calculate a flightpath and indicate the resolution map that will be created when the images are stitched together. (Courtesy Brandon Basso)

PROFESSIONAL CAMERA

This octocopter (8 motors) is able to lift a camera gimbal with a heavy SLR camera for high quality imaging.

DJI SPREADINGWINGS S1000

The bare frame of the DJI pro frame the Spreading Wings, which has retractable legs to allow for camera rotation in flight.

CHAPTER 02
POWERTRAIN

**All the pieces that get a
UAV in the air (and back)**

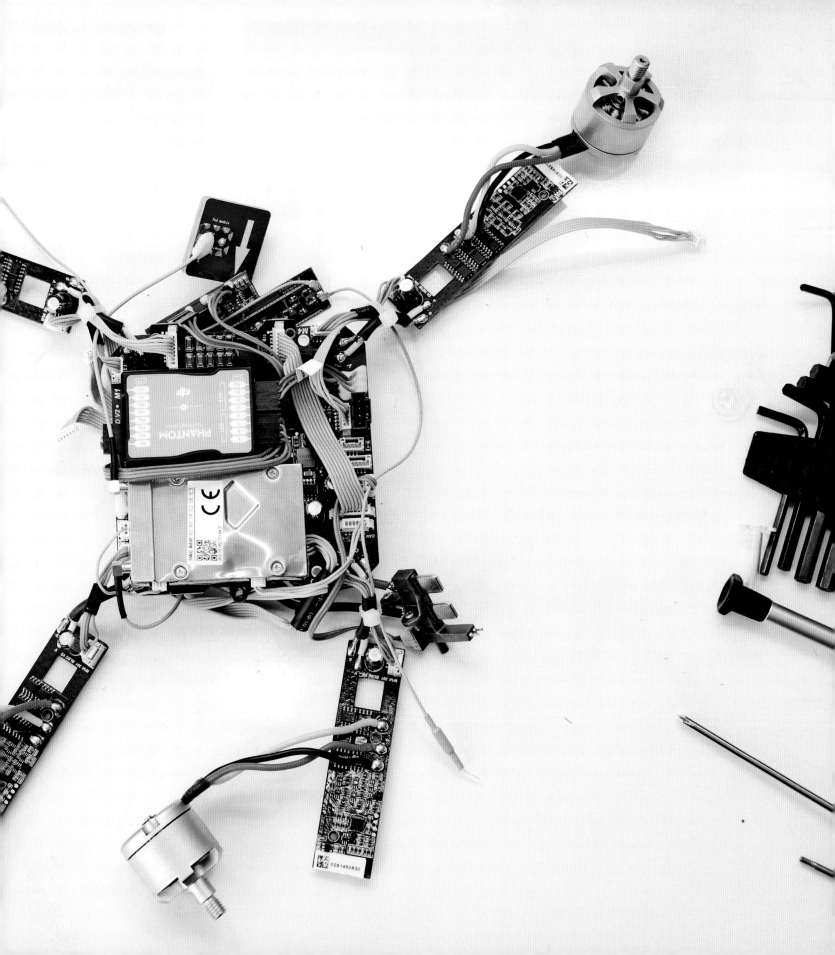

AIRFRAME

02

POWERTRAIN

In aviation parlance, the airframe is the mechanical structure of the aircraft, minus the propulsion and technical systems. In the world of multicopter construction the airframe is the structure that supports all of the other components, and there are some surprising alternatives to the four-rotor X-shape.

AT FIRST THAT meant that each multicopter was designed around a simple invisible circle with rotors placed evenly around the center of gravity, like the standard quadcopter, hexacopter or so on. The only real variation was whether the drone had one of the rotors directly at the 'front' or two either side of the front; though again this is more a matter of control.

In truth, however, though there is no real reason why the airframe has to feature equally long arms around a central piece. Given that, relatively speaking, the battery and payload tend to be very heavy parts of the craft then clearly the load on the motors should

be balanced, which means that the weight will end up in the middle, but since software — not aerodynamics — is keeping the trim stable, there are many other possibilities. Essentially so long as the control software understands where the props are in relation to each other the airframe can work. This makes possible less symmetrical frames, though unless you're willing to write new software for your flight controller you'll want to make sure you choose a supported one.

The classic 'X' shapes are certainly the most popular and with good reason; they don't directly block the forward view (should you mount a camera at the hub), offer the fewest challenges for design and making balancing easier. The Y6 and X6 designs (shown opposite) double the motors and props on each arm, which offers more thrust because the motors are doubled up and (with the mountings further apart) larger props can be used.

NLROBOTIC Y6 PRO
The Y6 has 6 rotors and 6 motors, just like a hexacopter. This is achieved, however, by putting two motors on each arm (rotating in opposite directions to prevent torque spin). The advantage of this arrangement is that there is a wide gap between any armatures allowing for wide angle photography without a prop in shot. (courtesy http://nlrobotic.com/)

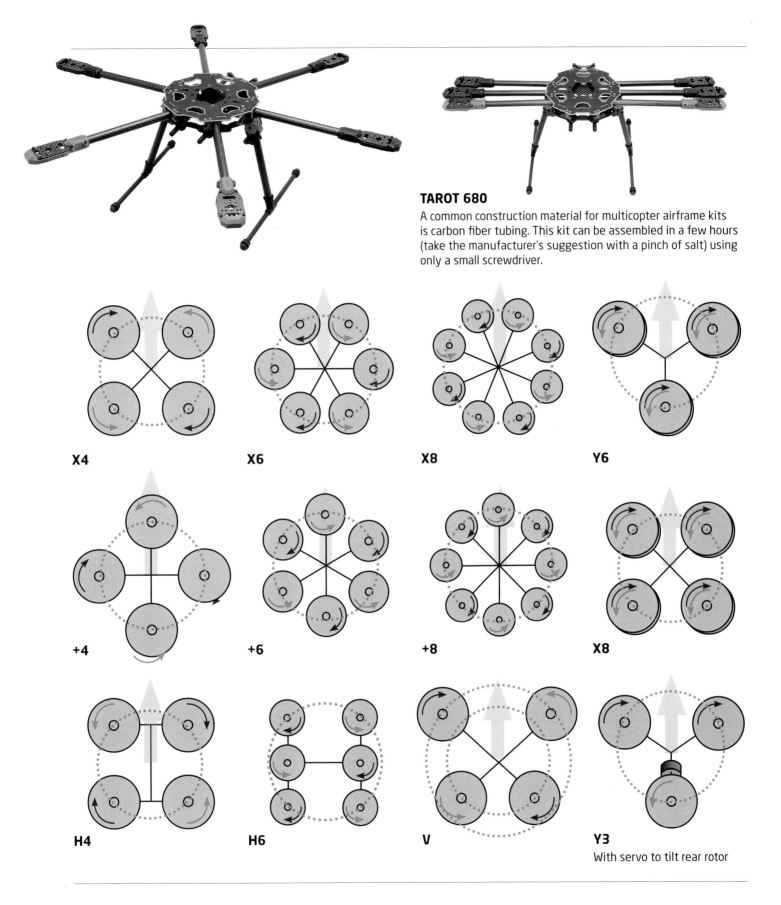

TAROT 680

A common construction material for multicopter airframe kits is carbon fiber tubing. This kit can be assembled in a few hours (take the manufacturer's suggestion with a pinch of salt) using only a small screwdriver.

X4

X6

X8

Y6

+4

+6

+8

X8

H4

H6

V

Y3
With servo to tilt rear rotor

BATTERIES

Lithium-Ion Polymer ("Li-Po") battery technology has helped make multicopters possible. These batteries are compact, comparatively light, and able to store a surprising amount of power. However, they can also be very dangerous: they need to be treated with respect, and you should know a little bit about them before you start using them.

WE'RE ALL ACCUSTOMED to recharging batteries, but the process is pretty well hidden—iPhone users never even see the batteries unless they break their phones apart, and charging is simply a matter of plugging in. RC pilots need to know more.

A Li-Po battery pack consists of one or more "cells," each producing a nominal 3.7 volts of power when it's running smoothly (although they can be charged to a maximum of 4.2v). Therefore, a three-cell battery pack would produce 11.1v (3.7v × 3). The shorthand for three cells is "3S".

The amount of power contained in the battery—the size of the tank, if you like—is typically measured in mAh (milliamp hours). Higher mAh ratings indicate a greater capacity, which leads to longer flights, although it also leads to heavier batteries, so you won't get as much more time as you'd like!

Finally, the discharge rate is the speed that the current can flow from the battery. It is common for this to be written as two numbers, such as 25C/35C. The lower number is the constant flow rate and the higher value is the "burst rate," which can be sustained for just a few seconds (for take off, for example).

Any battery with more than one cell needs to be charged using a balance charger that can monitor the state of each individual cell. This connection is called the balance plug or balance tap.

Capacity (mAh)
mAh represents the capacity of the tank (5000mAh, for example).

Volts
Voltage (for example 11.1v) is like the force pressing down on the tank.

The ESC (see page 38) acts as the tap.

Discharge
Discharge rate is equivalent to the width of the hose.

CELLS	VOLTAGE	CHARGE LIMIT
1S	3.7	4.2
2S	7.4	8.4
3S	11.1	12.6
4S	14.8	16.8
5S	18.5	21
6S	22.2	25.2
8S	29.6	33.6
10S	37.0	42

The voltages and charge limits for a number of battery cells. Even going a tiny way past the charge limit can destroy a battery pack.

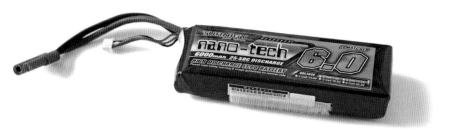

MULTI-CELL BATTERY

Any Li-Po battery with more than one cell has two sets of connectors. The larger connector provides the power and the smaller connector (the "balance plug" or "balance tap") links to the individual cells. This is required for balance charging (see below).

CONSUMER BATTERY: DJI

The Li-Po battery for a DJI Phantom has an easy-to-use case with a built-in power meter. The battery itself uses the same three-cell, Li-Po technology as the Parrot battery shown below.

BALANCE CHARGER

A balance charger needs to be told the type of battery (here it's Li-Po), its amperage (here 6000 mAh/6 Ah), and the voltage/number of cells (here 11.1v). The power connector and the balance tap are connected and charging commences. For a battery like this, a full charge would take approximately 30 minutes.

CONSUMER BATTERY: PARROT

The battery for Parrot's popular AR.Drone is a three-cell Li-Po battery. It's secondary (balance) charging points are built into the casing and marry up to pins on the supplied charger.

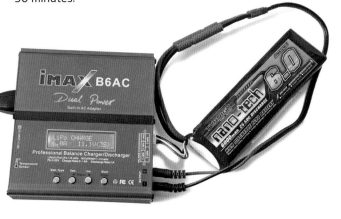

DANGER! SAFETY FIRST

Never leave Li-Po batteries unattended while they are charging, and once a battery is damaged do not attempt to charge it at all: they can quickly start fires (search YouTube for "LiPo Fire" for examples).

MOTORS

Spinning at around 8,000 revolutions per minute and able to very swiftly alter that speed according to instructions from the control system, multicopter motors must be very powerful. Given that the most common 'copter is the quad, there is also no redundancy; the motors must continue to turn to stay in the air, so reliability is just as essential.

THE INDUSTRY STANDARD is the brushless electric motor. Traditional brushed motors (a technology invented in the 1880s) suffer from power loss the faster they turn, not least because of the drag created by the conducting brush which was used to detect where the motor was in a turn and switch the power to maintain the turn. Brushless motors replace this approach with solid state circuitry that detects which direction current to apply and switches it without the need for brushes.

As well as eliminating the brush, the microprocessor control makes possible the rapid changes in speed needed by 'coptors; that's because each part of a turn is a separate calculation — indeed these motors are often called stepped motors in other applications because each partial rotation is a separately controlled micro-step.

As well as their accuracy, Brushless motors are very efficient, which is why you'll find them in applications from DVD drives and computer fans to devices like Segway people transporters and industrial robots. Their power to weight ratio, in combination with the LiPo battery's ability to deliver the required power, is what makes multicopters possible.

The only real issue is temperature; all brushless motors operate within strict temperature limits and efficiency drops off beyond limits, which its why, if you're building your own aircraft, it's crucial to select motors able to lift your planned weight.

The accepted standard is that your motors are not only able to lift the full weight of your 'copter (including all components and the batteries), but able to do so twice over. That's because just being able to lift the craft would create a perfect hover; you need the power to climb and some room for maneuverer to avoid burning things out. Three times makes things pretty sporty, but too much thrust will make things very hard to control.

Remember to divide the weight by the number of motors, so if your 'copter weights 200g then to get 400g worth of thrust you need 4 motors with 100g worth of thrust. Were it a hexacopter of the same weight then you'd only need 67g thrust from each motor.

Quadcopters are mostly Direct Drive (DD), meaning the prop is attached directly to the motor without any gearing, although the Parrot AR.Drone uses cogs. This adds an extra shaft to the mechanism and a common upgrade is to replace the fixed runners with bearings that reduce friction, but of course direct drive eliminates it altogether.

Brushed motors are common is small, cheaper copters, especially those for flying indoors. They have a limited lifespan — perhaps as little as ten hours flight time before they need replacement. In reality with a 5-minute limit on most such craft it may be adequate to see the pilot up to the point of mastering the controls and investing in something a little more robust.

HUSBAN MOTORS

These tiny brushed motors, with a diameter no bigger than a dime, are a replacement set for Husban's hand-sized X4 mini quadcopters.

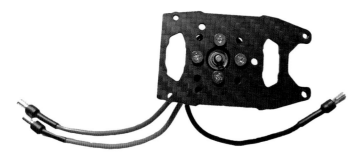

MOTOR FITTINGS

A selection of pre-drilled holes on the Tarot frame. In this picture a smaller motor is fitted than in the picture below.

FITTED MOTORS

Four pancake style "outrunner" motors (meaning that the outer shell of the motor rotates with it) attached to the pre-drilled platforms of the Tarot 690 airframe.

STANDARD FITTINGS

The underside of this well-ventilated SunnySky motor shows the four screw-fittings the manufacturer have provided to attach it to the aircraft. Note that each pair of holes are the same distance from the motor's center, but not the same as each other; two different options are supported.

KEY SPECIFICATIONS

KV The lower the kV rating, the larger the prop you're able to turn, and larger the prop the more thrust. 600–1200 kV is normal, below 600 kV are the most powerful.

Max current (A) The maximum current in amps the motor can handle at once. The battery provides the current, and the ESCs (see page 38) must also be able to handle it.

Suggested prop Motor manufacturers will clearly indicate which prop size & angle will work most effectively with it.

Shaft diameter Although most props come with adapters for various shafts.

Efficiency (grams of pull/W) There will be different results at 50% and 100% throttle.

Weight Each motor contributes to the overall craft weight.

ESC: ELECTRONIC SPEED CONTROLLER

The ESC stands between the low current required to operate microelectronic flight control systems and the raw power required to turn propellers. It reads the speed the motor is turning at, along with the information from the flight controller, and applies the power to keep the motor turning at the desired speed.

UNDERSTANDING EXACTLY HOW an ESC works is unimportant when constructing a multicopter, but you do need to know that you're choosing the right ones for your purpose. For a start, ESCs for brushed motors and brushless motors are entirely different animals, so if for some reason you're working with a brushed motor, you'll need to find a matching ESC.

Modern ESCs have their own control chip, which interprets the input signal from the flight controller and controls the motor accordingly. This interpretation comes in addition to the interpretation of your piloting decisions made by the flight controller in the first place. This interpretation might take the form of softening the rate at which the motor changes speed, for example, meaning that even if the flight controller sends the best possible signal, there will be a slight delay.

In the worst possible scenario this will make the 'copter impossible to fly, but it's far more likely that you will simply experience a softening of response times, especially when rapidly accelerating or decelerating. Initially, the solution was for hobbyists to upgrade the firmware on the ESC, a fiddly process that involving direct access to a microprocessor, rather than simply plugging in a USB cable (as you might when updating the firmware on a digital camera or other consumer electronics device).

An easier solution is to use appropriate firmware from the start. Mercifully, the market has provided a number of multicopter-friendly ESCs, which provide the snappy level of response that makes a pilot's life easier and exciting maneuvers that much easier to pull off. Perhaps the best regarded is the SimonK firmware (named for it's programmer).

However, while firmware is important, the most crucial measure of an ESC is the number of amps it can handle: it's vital that your ESCs are capable of handling the maximum load you will be running though them. In other words, if your motor draws 30A, you should ensure that the ESC you pair with it can handle at least that amount. It's always best to add a margin of error (at least 10%), so the for a 30A draw, the ESC would need to be rated at 33A (or, better, 35A).

Of course, you can plan for the future at this stage and go even higher, but as an ESC contains a heatsink to help dissipate the heat it generates, the higher the current the ESC is capable of handling, the heavier it becomes. Not only that, but the cost rises with capacity—given that you need an ESC for each motor, that can be a more significant factor than you might first think.

Finally, look out for a battery eliminator circuit (BEC or UBEC). These convert the reduce the battery's higher voltage down to a level that is suitable for powering the system's control electronics (thereby eliminating the need for separate, lower voltage batteries). For larger craft it's a good idea to make sure you use a separate BEC, as using the ESC's BEC will generate more heat.

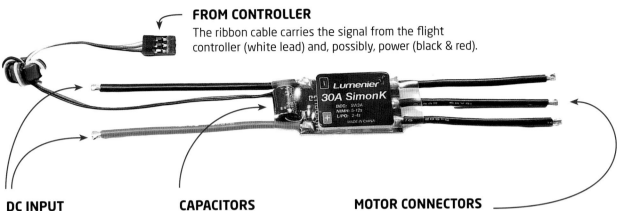

FROM CONTROLLER

The ribbon cable carries the signal from the flight controller (white lead) and, possibly, power (black & red).

DC INPUT

The thick red and black cables carry direct current (DC) from the battery (or the distribution board).

CAPACITORS

In order to avoid any feedback that might damage the battery, large capacitors cushion the DC flow (including any from an on-board BEC).

MOTOR CONNECTORS

Three leads carry alternating current (AC) from the ESC to the motor. Unlike the DC input these do not have poles—swapping so any two of these leads around will reverse the direction that the motor turns.

3D

3D Robotics' IRIS drone has a four-in-one ESC. The neat rows of chips on the back (left) send power to the motor. However, they also generate heat, so an aluminum heatsink is attached to them (right).

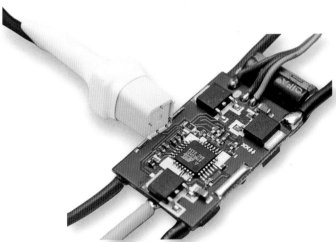

FIRMWARE FLASHING TOOL

The square chip seen on this uncovered ESC (the Atmel Atmega), is a common one, and there is a specialist tool that makes it easier to "flash the firmware" (update it). To do this, the tool's ribbon cable connects to a computer's USB port (via a *USBasp AVR* adaptor) and the tool itself clips onto the chip. The contacts allow the computer to write new firmware to the chip.

EMAX 4-IN-1 ESC

Specifically designed for quadcopters, the EMAX Multirotor 4-in-1 is just that—four ESCs with SimonK multicopter firmware in a single box. The elegant simplicity is revealed not only in the single red and black wires to the battery, but the minimalism of the ribbon cable to the flight controller: a single cable carries the red and black power lines.

PROPELLERS

Selecting your motor is only part of the lift equation: you also need to know what kind of propeller, or "prop," it will be turning so you know how much lift it will be generating. Props are sold in a variety of different materials, all of which have some impact on their effectiveness, but the key factors are size and pitch.

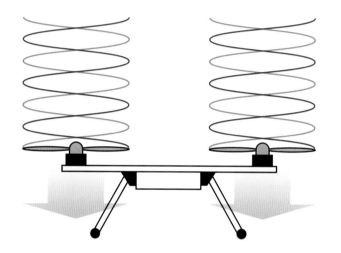
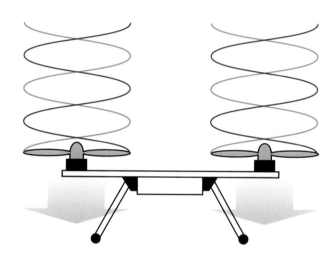

ALTHOUGH MULTICOPTERS LARGELY live in a metric world, the most common measurement of a prop is given in inches, in the form "10 x 4.5" (or similar). Here, 10 is the diameter of the imaginary disk that the propeller creates when it spins, and 4.5 is the pitch, which is a measure of the distance the propeller should travel in one revolution of the engine.

The maximum length of propeller you can use is normally dictated by your airframe, as you don't want them to overlap.

Assuming they're the same length, a lower pitch propeller will need to spin faster to generate the same lift as a larger one. That doesn't mean you should simply aim for the highest pitch available, though, as higher pitches tend to generate more

turbulence: if your 'copter is unstable, this might be something to look at. Luckily there isn't much of a market for props that make all craft impossible to fly, so you'll find it very hard to specify a truly useless prop. If you do have instability, it is more likely that your propellers would benefit from balancing, which is discussed later (see page 130).

Motors have an optimum speed at which they're most efficient, so to stay in the air longest you'll need to ensure you've selected a propeller that rotates at a speed consistent with your motor's comfort zone. Again, this is covered in more detail later on.

Smaller propellers are easier on 'copter motors because they generate less inertial movement than larger ones,

CARBON FIBER PROP PAIR

Quadcopter props are sold in pairs: one to turn clockwise, another to turn counterclockwise.

making them easier to get started. They are also less likely to continue rotating undesirably when the motor adjusts its speed (as it needs to do continuously to maintain controlled flight or hover).

One of the great things about props is that you can change them relatively inexpensively and easily, assuming there is suitable leeway in your motor and ESCs. There are several different materials that props are commonly made of, including plastic, carbon fiber, and wood. The latter is less common—especially in off-the-shelf products—but you will find wooden props at model shops.

It is recommended that you start out with cheap, disposable plastic props—these might even be all

you ever need. That said, you can gain thrust using more rigid carbon fiber props, so for applications where lift and stability is more important than speed (carrying a heavy camera, for example) they do have an advantage.

On the downside, carbon props—especially those that have come straight from the supplier—can be dangerously sharp. This isn't an aerodynamic advantage, but an outcome of the manufacturing process, and one that adds risk. They're also more expensive to replace—and the first lesson most multicopter pilots learn is that they'll need some spare props.

PROP ADAPTORS

Propellers are either attached directly to a multicopter motor or clamped into place using a prop adaptor. The adaptor has a hole in the dome for tightening.

DANGER! SAFETY FIRST

Never exceed your frame's maximum propeller length. Each prop must have complete freedom to rotate within its own diameter (although the gaps between them can be quite narrow).

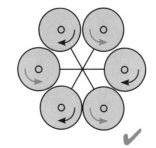

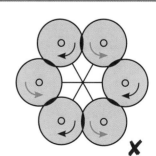

FLIGHT CONTROLLERS

In simple terms, the flight controller is your 'copter's brain. It's the central hub where information from your 'copter's attitude and position sensors and commands from your RC controller are received and from which instructions sent to the motors.

IN PRACTICE, THE flight controller is also likely to be home to a number of the sensors that it relies on for its information. That's because more and more multicopter-specific controllers are emerging, which often include gyros and accelerometers.

Two different philosophies of flight controller have emerged amongst the community: the closed-system (epitomized by DJI's enthusiast level NAZA and professional A2 flight controllers) and open source community projects (such as OpenPilot). Somewhere in between stands 3D Robotics, a commercial company that is symbiotic with its open source community, DIYDrones.

In all cases, the device is a small computer system with a processor and some sensor chips on a small board. The presence of the sensor chips means that it's position on the airframe will likely be important— it will usually need to be close to the center of gravity and pointing forward. It may also have external sensors included with it, such as a GPS unit.

Aside from the sensors, the key input to the flight controller is the signal from each channel of your radio control. For this you'll use the a receiver paired with the radio controller. Traditionally, these have featured a single signal wire for each channel, so you would have at least five cables running from a bank of pins on the receiver to their corresponding pins on the flight controller's inputs (although these are typically three-wire cables with ground, power, and signal wires). Some RC receivers feature PPM which can dispense with the individual wires.

DJI NAZA V2

A popular and proprietary flight controller, the DJI NAZA is compact, with one end featuring a bank of 3-pin connectors which connect to the radio receiver as well as two DJI-only 4-pin connectors for the accessories. At the other end a bank of 3-pin sockets connect to the ESCs.

APM 2.6

Sold through 3D Robotics, this Ardupilot/ArduCopter-based flight controller is popular amongst hobbyists thanks in large part to its compatibility with the sophisticated and relatively accessible AMPlanner software.

SENSOR	HOW IT WORKS	ESSENTIAL?
Gyroscope	A gyroscope detects it's own (and hence the 'copter's) orientation with reference to the Earth's gravity. Essentially it is the digital equivalent of the artificial horizon.	Yes
Accelerometer	An accelerometer measures non-gravitational acceleration. At its core it contains microscopic crystals that become stressed when vibrated and release measurable voltage.	Yes
Barometer	A barometer is used to measure altitude; air pressure is lower the further up you fly, and by measuring the change the barometer can be used to determine the altitude. Accuracy can be affected by terrain and any weather. The stated barometric altitude is based on an unchanging average sea level, not the local level. Can be used for the flight controller to maintain altitude.	For Altitude Hold mode
Magnetometer	Essentially a compass, this device measures the direction of a magnetic field. The 'copter can use this to measure where magnetic north is, which helps with orientation (and is essential if GPS is being used).	For Loiter (Position Hold) mode, Return to Home
Optical Flow and SONAR	A sonar (echo location) can measure the distance-to-ground when you're within a few meters of it, making it more useful than a barometer for landing and near-ground flight. Optical Flow tracks objects on the ground using a downward facing camera.	For some auto-land functions.

02

POWERTRAIN

Figuratively (and often literally, depending on the design) the other end of the controller is the signal output. Once again you will typically find a bank of three-pin connectors (in this case for the ESCs), carrying a signal (white/yellow), power (red), and ground (black/brown) wire.

Your choice of flight controller is determined by your budget, the kind of flying you'd like to do, and your desire to get involved in a community or to buy products from a reliable commercial entity. Those with cameras on their minds should also consider whether the flight controller offers any compatibility with their chosen camera-mounting system, as the NAZA and Zenmuse do.

Another useful feature is the ability to output telemetry data (live flight information from the sensors, such as the current altitude, heading, or battery voltage). The latter can be sent to a dedicated receiver or super-imposed onto a video feed which is transmitted back to the pilot's display (or video goggles).

KK MULTICOPTER
Processor: Atmel ATmega644A (8-bit, 1MIPS)
Sensors: 3 gyro & 3 accelerometers

This is a great first board for DIY droners, featuring as it does an on-board LCD for setup so no computer is needed. Performance is nothing special, but it's cheap and allows you to tweak gains in the field without plugging in a laptop, so if you're building a first quad for fun it's a great choice. The KK project takes its name from the programmer who began it, KapteinKuk (Rolf R. Bakke), and he writes the firmware entirely in assembly language. This is the most machine-friendly form of computer code, but also the hardest to write, so there are not many collaborators developing it.

Signal

MULTIWII PRO
Processor: Atmel ATmega2560 (8-bit, 16MIPS)
Sensors: 3 gyro, 3 accelerometers, barometer, magnetometer and optional GPS.

SERVO CABLE
You'll typically connect your flight controller to the radio receiver and to the motors with servo cables. This one is female to male and would connect to male pins the board. Where there is a user-friendly case like the DJI NAZA the connector should be designed to only fit one way round, in which case follow the wiring diagram.

The MultiWii Pro is Arduino based and features a full range of on board sensors as well as a USB connector which can be connected to a computer for programming. Optional LCD displays are available for field configuration, or a you could choose a bluetooth daughter-board. The MultiWii was originally designed to take advantage of the cheap sensors available in a Nintendo Wii nunchuck controller, but now is a fully-fledged flight controller with active developers looking forward to autopilot control.

APM2 (MEGA 2)

Processor: Atmel ATmega2560 (8-bit, 16MIPS)

Sensors: 3 gyro, 3 accelerometers, barometer

Supports: Magnetometer, GPS

The big advantage of the APM is that it grants access to the powerful and accessible APM software. It's also a powerful flight controller, using co-processors to handle USB connectivity.

The ArduPilot began life as a fixed-wing control system that grew to include rovers, helicopters & multicopters (as ArduCopter). It can be developed using Arduino, Windows Visual Studio or Linux, and is run as 3DRobotics's open-source project.

PIXHAWK

Processor: STM32F427 (32-bit, 225MIPS)

Sensors: 3 gyro, 3 accelerometers, barometer

Supports: Magnetometer, GPS, sonar, Telemetry

The pixhawk is a powerful system capable of broadcasting telemetry and/or recording it to its on-board MicroSD card. There are a variety of control systems available, including full autopilot, as well as an active community, and Pixhawk uses the same GUI-based mission software as the APM.

The Pixhawk was developed by 3DRobotics which was looking to expand its work on the APM but was running out of processor power, so took the code they already had and ported it to a new system.

OPENPILOT

Processor (original): STM32F103 Arm Cortex M3 (32-bit, 90MIPS)

Processor (Revo): STM32F405RGT6 32bit ARM Cortex M4 (32-bit, 210MIPS)

Sensors: 3 gyro, 3 accelerometers (Revo adds Barometer, Magnetometer)

Supports: Telemetry, GPS

OpenPilot have been going a long time and have two main systems out there, the CopterControl 3D (CC3D) and the newer Revo. board. In both cases the development if community based (with the advantages and disadvantages that brings), and the newer version brings significant processing port for the team to add powerful new features.

DJI

Processor: [Various]

Sensors: 3 gyro, 3 accelerometers, barometer, Magnetometer, GPS,

DJI's range of flight controllers begins with the elegant NAZA Lite kit, and offers built-in features to stabilize DJI's 'Zenmuse' range of camera gimbals very popular with photographers; further up the range additional DJI photography, telemetry and live HD video relay features (including app-based GPS waypoint setting) become available and they require remarkably little setup.

Although their rapid growth of the ready-to-fly market has left some hobbyists feeling DJI are a bit too corporate for them, that's a personal choice and the equipment has much to offer.

RC TRANSMITTERS

POWERTRAIN

Although not every drone is controlled by an RC-unit, some pilots opting for wi-fi devices like Parrot's offerings, and others choosing fully automated flights, the RC-unit remains the core of pilot-drone interaction. It's a busy marketplace, and multicopters come with certain requirements.

YOUR RC TRANSMITTER IS a real confluence of jargon; not only do you encounter the different control modes (1 through 4), but even their explanation is usually described in terms of the throttle and traditional control surfaces of a traditional aeroplane: rudder, elevators and ailerons rather than the direction your copter will actually move in, but here we'll think of it in 'copter terms.

There are a few crucial things you'll need to look for when selecting a transmitter (or Tx as it is often abbreviated to); firstly the number of channels. As discussed early on, each channel can influence one aspect of flight; so four channels are needed for the basic 'copter controls: vertical (throttle), rotate (rudder), forward/backward (elevator) & sideways movement (aileron).

Additionally you'll need at least one channel for a flight mode switch; any of the physical switches can be assigned a channel using the menus on a controller like the FR Sky illustrated and this can be used to switch between flight modes like "Altitude hold" and "GPS hold". Other channels might be configured for additional features on your copter, like a link to a servo that can tilt your camera.

When you're choosing a controller you'll also need to choose a control mode. That's because, unlike the other flight control channels, the throttle's thumbstick is not sprung; it can be left anywhere from minimum to maximum. That means a Mode 1 (throttle on the right) controller is physically different from a Mode 2 (throttle on left) one. Both modes have secondary arrangements; 1&3 and 2&4 use the same hardware.

FR SKY TARANIS

The FR Taranis is a popular choice amongst experienced pilots, offering as it does up to 16 channels. Note that the Mode 2 configuration – the most common in the US – is shown above and the Mode 1 to the right; the difference is that the throttle can be left i n the zero position while the other main controls are sprung back to center. The different arrangement of physical parts can be swapped around manually (with a screwdriver).

If you've never flown before, it's probably best to go with the most popular and these days that's almost exclusively Mode 2. This was always popular in the US, though elsewhere in the world Mode 1 was more common amongst hobbyists, but since many people's first experience is now a ready-to-fly product (like a DJI phantom) in Mode 2 it seems sensible to stick with it. If you find you really would prefer Mode 1/3, some transmitters can be opened up so the switches can be swapped. There are a number of other Tx features to look out for; the kind of signal it sends (some send all the analog channels in a combined digital signal), and whether data is relayed back (see page 50).

CONTROL MODES

There are two possible physical arrangements giving rise to four possible control modes.

Throttle   Rudder
Elevator   Aileron

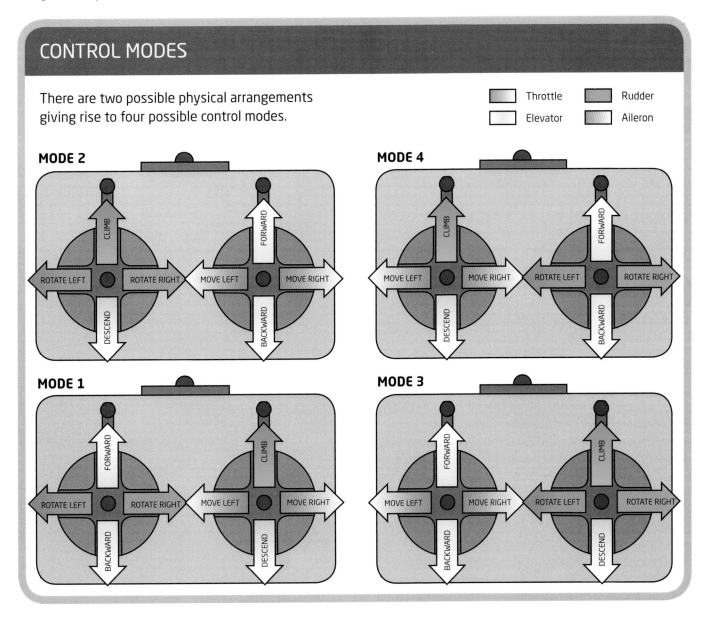

MODE 2

MODE 4

MODE 1

MODE 3

FPV SYSTEMS

First Person View, or FPV, is a complete hobby within a hobby when it comes to multicopters. A camera sitting in the "pilot's seat" relays video via a transmitter to the pilot who can see the drones-eye-view on a screen or, as is more popular when racing, on video goggles.

USING A DIFFERENT radio frequency from your RC-controller its possible to send video back from an on-board camera to a display or video goggles. Indeed the camera and transmitter are so relatively inexpensive that they are frequently bundled with the video goggles.

The key to making it work is what initially feels like slightly old technology; an analog PAL or NTSC video signal (which has none of the encoding and decoding lag of digital) is transmitted from the craft to be picked up by anyone watching that frequency. Just like TV there are different channels and its vital only one drone is broadcasting on that frequency, hence the requirement for everyone at a meet to be on the ground when powering up — if someone turns on a drone using your frequency while you're flying FPV you'll lose your signal and, in all likelihood, crash. (This is exactly the kind of thing the frequency hopping system on the main control frequencies was designed to avoid.)

It's vital that the radio system is omni-directional, and clover-leaf antennas are far better at achieving this than the simple whip antennas (sticks) included with most kits. Antennas are typically sold polarized to their particular frequency to cut down on interference so it's vital to buy one for your frequency group, probably 5.8GHz.

As you turn quickly in and out of the sun, especially when flying for sport, the resolution of the FPV camera is less important than its ability to quickly adapt to the exposure circumstances; in other words you want to be able to see detail whether you're flying with the sun behind you or straight into it. Your best bet is to check fellow pilots videos online before selecting a camera.

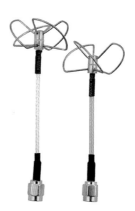

OMNI-DIRECTIONAL ANTENNAS
This Emax pair of 3 and 4 leaf 5.8GHz antennas are polarized specifically for their frequency. The 3-leaf one is for the 'copter, the four-leaf for the pilot's glasses or monitor.

HD 700TVL MINI FPV
Selling for about the price of two coffees with all the trimmings, this FPV camera can easily be mounted on most copters and linked to a transmitter; it relays video using the same standard that older analog TVs used.

CAMERA ANGLE

This quadcopter is a classic FPV racer; the two boards of the frame protect the electronics from the inevitable crashes and the FPB camera has been angled so that when the 'copter is flying forward fast the video is level.

CHANNEL	FREQUENCY
1	5705
2	5685
3	5665
4	5645
5	5885
6	5905
7	5925
8	5945

FPV TRANSMITTER

This Boscam TS351 can transmit 5.8GHz video using a choice of 8 channels. Note that there are tiny switches under the plastic wrap at the top left; these can be adjusted to change the frequency according to a the manufacturer's instructions (it works out in binary fashion: 0000 for ch 1, 1000 for ch 2, 0100 for ch3 etc.)

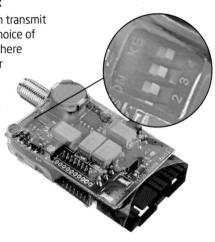

FPV GOGGLES

By far the dominant brand is 'FatShark',

TELEMETRY

As well as video, it's possible to get back—either live or by recording—all kinds of other information, like battery status, altitude, heading and so on. Collectively this information is known as telemetry and making use of it can really change the way you fly, or just help you geek out.

DEPENDING ON YOUR choice of flight controller, a great deal of the information that counts as core telemetry is already known to the system keeping your drone in the air. Getting it back to you is a simple matter of deciding how to transmit it. If you're interested purely in reviewing your flight data after-the-fact then you'll need to look for a flight controller like the Pixhawk which has the ability to log telemetry. The log can be used by manufacturers, or fellow fliers, to help you diagnose issues.

If, however, you're more interested in seeing the information as you're flying then you'll need to transmit it somehow. The DJI Phantom Vision, for example, employs a Wi-Fi range extender which allows it to transmit the telemetry that it's flight controller detects as well as the video signal, hence the bulky additional lump (with separate batteries on earlier models). Wi-Fi has the advantage of being a two-way system, so can also be used for additional controls.

With 'copters like the Parrot drones, where flight control is also via Wi-Fi, the telemetry is added as a matter of course and the App used to control the craft can also include it when you upload the video to the in-built community afterward.

When it comes to self-builds, you first need to know what data is available, and from where. The flight controller is the hub of all the data, but is it prepared to share? Only a few years ago it was common to buy a telemetry hub which acted a little like a flight controller but simply converted the data so it could

be transmitted. That could require a duplicate set of sensors. Now flight controllers like Arducopter 2.4, Pixhawk and the Naza have telemetry-out which will, with the help of an adapter.

Telemetry systems create a two-way radio system, and in many cases this is a separate one to the main RC control. Why not simply use that link to control the copter? RC is very, very reliable, and serial links which have to carry many packets of data are considerably less so, meaning that while you can send commands from a laptop if you choose — some even connect joysticks to their PCs — you should always have the RC to hand.

TELEMETRY PORT
The Arducopter flight controller has a dedicated telemetry port which can be connected to either a bluetooth link (which only works at short range but is nevertheless useful for calibration commands) or a radio serial link.

WI-FI TELEMETRY
Information like altitude, seen on the DJI Phantom 2 Vision + display.

IOSD DISPLAY
Here the iOSD display as it appears overlaid on a video signal; the diamond points to the pilot's location and artificial horizon is laid over the center.

Altitude (GPS) (m) 0.00	Altitude (REL) (m) 0.00
Battery (%) 0.00	Climb (m/s) 0.00
Current (A) 0.00	GPS Fix () 0.00
GPS HDOP (m) 0.00	GPS Sats () 0.00
Pitch (deg) 0.00	Roll (deg) 0.00
Voltage (V) 0.00	Yaw (deg) 0.00

LAPTOP TELEMETRY
The tool APM displays telemetry data and an artificial horizon using data from a radio serial link.

RADIO SERIAL LINK
This is a pair of modems connected with radio allowing for not only telemetry but working just like a direct connection to a computer or tablet for setup.

2-WAY RC
Some Radio Control systems, like those offered by FrSky, support telemetry with two-way radio. The 'receiver' unit for the craft actually features two antenna, one for the telemetry transmission.

2-WAY RC
At the receiver end, a matching module plugs into the RC unit and the display screen on it relays important information including the especially vital battery voltage.

IOSD MINI
DJI's flight controllers support (to varying degrees) a system called iOSD which can take data from the flight controller and display it on the video of an FPV system.

LANDING GEAR

Traditional aircraft retract their landing gear for a more aerodynamic shape to allow for faster, more efficient flight. 'Copters have little to gain in this department so at the bottom end of the scale landing legs can be pretty basic. Retractable systems, or 'retracts', instead become a popular extra with photographers.

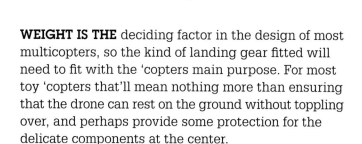

RETRACTS FOR SELF-BUILD

Retractable landing skids for self-builds are available, with degrees of control that vary widely, along with the price.

BUILT-IN RETRACTS

DJI's popular Inspire 1 features a built in retracting system that automatically lowers the legs for landing as it approaches landing height, then raises them out of the way of the 360° camera

WEIGHT IS THE deciding factor in the design of most multicopters, so the kind of landing gear fitted will need to fit with the 'copters main purpose. For most toy 'copters that'll mean nothing more than ensuring that the drone can rest on the ground without toppling over, and perhaps provide some protection for the delicate components at the center.

Even fairly sophisticated craft need nothing more than that; your author's Tarot 690 frame hexacopter features nothing more than two upside-down T-shapes as legs, yet these never get in the way of the camera at normal angles. My camera, however, always points forward. Where the camera has the freedom to rotate, as on most professional rigs the extra weight to add a servo to retract the legs is well worth it; camera operators can turn around the central hub to keep on a subject however the pilot has chosen to fly.

POWERTRAIN

02

ECLIPSE

This shot of a DJI S1000 was captured by Peter Haygarth of Droneflight, a UK based aerial imaging firm during the 2015 solar eclipse. The bottoms of the retractable legs of the 'spreading wings' craft are visible to the left and right of the ten-rotor drone.

With ready-to-fly aircraft like the Inspire the control mechanism for the gear is typically an electronic signal sent via the control link rather than requiring a separate channel, but for self builds without the luxury of a two-way digital link like Lightbridge or range-extended wi-fi a separate channel on the RC will need to be configured to switch the gear.

UPGRADES

Phantom pilots might be interested to know that the after-market has a number of solutions to add retracts to the original design which can be found from stores like Hobby King.

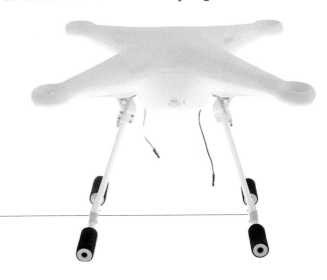

CHAPTER 03
ACE PILOT

Get off the ground safely, then master the skies. Goggles and leather hat optional

FLIGHT BASICS

All multicopters are generally flown in the same way, but just as a sports car handles very differently from a London bus, one multicopter might have very different characteristics to another. However, before you can discuss the feel, you need to know the language.

FLIGHT MODES

The transmitter's stick mode dictates which finger you'll be using for which maneuver, and the point at which you buy your transmitter is usually the point at which you would set its mode. Your 'copter's flight mode, however, is something you will often change several times in a single flight.

Flight modes tell the flight controller how to take advantage (or not) of its on-board sensors. For example you could choose the very useful Altitude Hold mode, which tells the 'copter to maintain its exact altitude when you release the throttle stick (which, as you'll recall, is the only stick that springs to center or 50% position). Alternatively, you might choose GPS Hold mode, which does the same thing, but maintains the 'copter's position in 3D space using satellite positioning.

With some ready-to-fly 'copters there may be no mode choices, or you might have a switch with some pre-determined options. If you're building your own system you will find that you have many more options (depending on the flight controller and system). In any case, ensure you've selected the correct mode before take off.

RETURN TO HOME

Most craft fitted with GPS offer a "return-to-home" mode, which means the 'copter will return to its take-off location if the connection to the controller is interrupted. If you have this feature, make sure

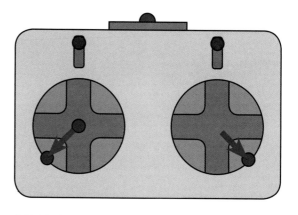

ARMING POSITION
Make sure to check your manual to see which stick position(s) will arm your 'copter's motors.

the 'copter has set the home location before take off, or you might find that it instead flies to the last point it remembers—possibly from a previous flight at an entirely different location.

ARMING

As a safety precaution to prevent accidental takeoff, a 'copter's motors generally need to readied to accept commands. This is slightly over-dramatically known as "arming," and is usually performed by holding one or both of the controller's sticks at the lower corners. Arming is the last thing to do before take off and disarming is the first thing to after landing.

FACE 'COPTER AWAY

When you take off, it's wise to have your 'copter facing away from you, so that . You should also stand far enough back to allow for the 'copter to drift a little at take off—at least ten feet, more if you or the 'copter are new to flying.

TAKE OFF

Some manufacturers include so-called "easy" take-off modes that automatically lift the drone off the ground then hold position. In truth, there is little difficultly involved in take off, especially with any 'copter capable of position hold—simply move the throttle stick up confidently. It's once you're off the ground that piloting can start to get more taxing.

FLIGHT

How you fly your 'copter will depend a great deal on the mode you choose. A GPS-hold/loiter mode will make it easier to get to grips with the controls at a comfortable pace; you can get your 'copter in the air to a reasonable height with the throttle, then push forward and practice adding turns and rotation.

LANDING

Unless your 'copter is fitted with a "land" button, the best way to land is to be decisive; identify a spot and

FLAT TAKE OFF SURFACE

Taking off from (and landing on) a flat surface is important. You should also make sure grass or other vegetation isn't interfering with either your props or your camera lens.

hover over it, compensating for the wind gently with pitch and roll (using the right stick if your remote is set to the standard, Mode 2), then pull the throttle down and commit fairly firmly to overcome air turbulence near the ground.

PRE-FLIGHT CHECKS

03

ACE PILOT

Before taking control of an airliner, a pilot is required to walk around their aircraft to make sure all is OK (and even if the authorities didn't mandate it, they'd be pretty crazy not to). Your drone might have cost somewhat less than that airliner, but you're both mechanic and pilot, so you'll have two sets of checks to perform.

IT SHOULD GO without saying that the point of checks is to avoid crashes. If something is going to go wrong with your 'copter—other than pilot error—it tends to happen very fast. A hexcopter or octocopter can survive the loss of a prop, but with a quadcopter, the loss of a single motor, prop, or ESC will usually be enough to have your beloved drone plummet from the sky and hit the ground with all the grace of a brick.

Even the most robust of 'copters will have its weak-spots, points at which the frame will fail. DJI Phantoms tend to fail just inside of the motor housings, for example, and your author has made quite a mess of a couple of Tarot 680-like frames with impacts from relatively low heights that still proved hard enough to fracture carbon fiber tubes as if they were ornamental glass.

BATTERY TESTER

This inexpensive device plugs directly onto a LiPi battery charging cable (the more cells the battery, the more pins you use). When connected, it displays the battery's voltage then cycles through the individual cells.

Repairs are not only expensive, but can also be frustrating. The self-builds that I fly use components from the Far East, and even though I have a great local supplier, who stocks a lot of the original airframes, they don't carry all the replacement parts. So, in the event of a heavy crash, I can find myself left with the awkward choice of replacing the entire frame or waiting for components to ship.

Finally, of course, crashing in public because you failed to check something obvious will not only hurt them and make you feel stupid, but it may harm the reputation of other fliers. Unfortunately, there is plenty of paranoia about drones at the moment, so you don't want to be the cause of another sensationalist "drones are dangerous" report, or the motivation behind an opportunistic politician seeking needlessly restrictive legislation. You shouldn't let that scare you into abandoning flying, but always be aware that the advantages of ensuring your 'copter is flight-safe extend well beyond saving you money on spare parts.

AT HOME CHECKLIST

Weather
Before devoting too much time to flight prep, it's worth making sure there's a chance of decent weather.

Prepare batteries
Check all the batteries you want to fly with are fully charged, then fit (but don't connect) one to your craft. Also check the batteries for cameras, remote controls, FPV goggles, and range extenders.

Prepare memory cards
Make sure any memory cards you'll be using in your camera have plenty of space on them and then put them in your camera/'copter so you are sure not to forget them.

Center of gravity
Place all equipment on your craft and ensure that the center of gravity is where it should be, so no single prop and motor bears too much load.

Mapping
Check the area you're planning to fly for any restrictions/airports. If you'll be using flight-planning software (such as APM), pre-load map tiles for the location you'll be flying.

Structural
Make sure that all screws are tight. Check that you've repaired any damage from previous flights and that those repairs are holding up.

ON-SITE CHECKLIST

Unpack slowly
Give everything a good look over as you get it out to make sure it's survived the journey to your flying location.

Tighten props
Check that all your props are correctly tightened, whether you twist them on at-site or in advance.

Clean lens
Make sure all the cameras fitted to your device (FPV, recording) are clean and ready to go.

Ground control
Set up and power up any ground-control systems, such as APM Flight Control.

Power up radio control
Make sure you power up the controller before your craft so you're ready to correct any problems.

Check video channel is clear
If you're using FPV video, check no one is flying on the same channel nearby (if in doubt, wait).

Power up craft
Turn on your 'copter, and if necessary, arm it locally (some 'copters offer both on-craft and remote arming).

Pets and children
Make a final check for dangers before arming and taking off.

CASE STUDY: SKYPAN INTERNATIONAL

UAVs have revolutionized aerial photography, and many people getting into flying now aspire to shoot photographs from the air for the property market. Mark Segal, and teammates Robert Harshman, and Jeff Jones pioneered this kind of photography with their company, SkyPan International.

MARK'S PASSION FOR aerial photograph extends back to his childhood, when at the age of five he "went up in helicopter to change Hasselblad film holders while my father, Ed, was photographing projects in Washington, DC." Mark says that he has "always loved photography, particularly panoramics, and started with Cirkut cameras at age 12. I assisted father on panoramic group assignments (which were his staple for 50 years) throughout the DC area before moving to Chicago in 1984."

Four years later, Mark formed SkyPan International, and the company now employs 7 people—a team of "seasoned professionals, all in [their] late 50's and each with over 25 years in their various specialties." These skills include robotic piloting/engineering, panoramic photography production/planning, and post digital wizardry experts, with additional freelancers called in when necessary.

The company's roots were in "rigging full scale manned helicopters with panoramic film cameras," and when they need to work above 500 feet, a manned helicopter is still preferable. Much of Mark's work involves mid-rises, not high-rises, though, and it isn't possible to use a full-size manned helicopter to fly below 500 feet in urban areas safely. The team "didn't like the lack of control with other systems, such as tethered balloons, blimps, or aerostats," which is where drones entered the picture. However, Mark is reluctant to use the term "drones," preferring instead to refer to the craft as either "robots, Remotely Piloted Vehicles, or RPVs—drones began as an early

SKYPAN TEAM
Mark Segal,
Robert Harshman,
Jeff Jones.

description for fixed wing defense/military aircraft ... it is not our terminology of choice." He also notes that even UAV has evolved into UAS (Unmanned Aerial Systems) to cover broader description by FAA.

Rather than looking for off-the-shelf aerial solutions, SkyPan International carefully builds its equipment based on its own—often very specific—needs. This has resulted in an array of "custom-made gimbals, camera mounts, and robots crafted from years of experimentation and under constant improvement." The technology also includes systems that use multiple camera mounts to capture 360° imagery that can be experienced online. However, the process is far from straightforward, requiring "five different software programs to assemble, retouch, and output superior quality imagery."

As leading users of remotely piloted vehicles, SkyPan International has also helped the FAA draft emerging

rules that will govern the commercial use of drones in the US. "New pilots need to gain experience by accumulating hundreds of flying hours before applying to commercial operations," says Mark. "Multicopters are anything but toys, and while most are well engineered and far better than the RC options in the 90s, they can still easily damage or even kill a person. No flights should operate over people, no matter what size." Yet despite safety concerns he still sees "unlimited potential and many unforeseen uses" for UAVs, making it "a very exciting time to learn and invent."

SKYPAN 3D IMAGE
Mark's team make interactive 360° images; here, for example, you can turn a virtual flying camera at the WTC1 construction site.

CASE STUDY: HELICOPTER GIRLS

With any hobby that is perceived as male dominated, you can be sure a marketing executive will want to "double their market" by getting women involved. Typically, that will involve releasing a pink version of the product and vodcasting women using it, but that's not remotely how the Helicopter Girls got started.

THERE'S NO HIT of pink on Natalie Samson and Emma Bowen's professional website, and they have also been established longer than many of their counterparts with Y chromosomes—their name acknowledges their gender simply because "when we started to show up on location with the gear, people started to call us the Helicopter Girls".

The Helicopter Girls started "because Em wanted to use an aerial shot in a documentary," and the pair haven't looked back. "We're pretty much the Helicopter Girls all the time now, but we've also worked on a lot of TV shows, producing for Nigel Slater [a UK TV chef], for example."

Working professionally involves a lot of paperwork, with risk assessments not only needed for the Civil Aviation Authority (the UK's equivalent of the FAA), but also for their clients—the paperwork generated by BBC Compliance, as the famous broadcaster's rule-checkers are known, is legendary.

The costs involved are also high—"When you've got a payload up in the air worth $200,000, the stakes change a bit (and so does the insurance)" Nat points out. Much of the cost comes down to getting the best-quality footage using high-end camera equipment. In the early days, equipment concerns were less important—"everyone was so dazzled by getting a shot from the air that people didn't really care about the quality of it"—but now the industry is already looking for more.

Keeping up with the technology hasn't been easy though, and Nat admits they haven't always made the right choices with their equipment—"I suppose these days you can probably get into it for about a tenth of what we've spent on equipment."

When they work, Nat is primarily the pilot, but Em also has her B-NUC qualification (the British certificate that allows commercial flying). However, while the Helicopter Girls was started as a commercial venture, it seems that its hard not to completely fall in love with flying—"one thing I'd not done until recently is FPV," says Emma, so watch out racers!

WORKING WITH A CLIENT
Em and Nat working between their clients.

GREENLAND
A music video shoot with the DJI Spreading Wings.

X8 DRONE WITH CINESTAR LENS
Eight large motors are needed to lift a cinema-grade camera.

BBC QUALITY
This subtly lit show was caught for the BBC and meets the company's stringent image quality standards.

FLIGHT MODES

03

ACE PILOT

Anyone who loves cars will talk about the handling of one vehicle versus another, or even the change the "sport" button makes on their own car. That's nothing compared to the changes you can achieve by changing flight mode.

IF YOU'RE RACING your 'copter around a tight path and want to be able to accelerate like a scalded cat, climb like a rocket, drop like a stone, and lean into turns for extra push out of them, then you'll want the flight mode set for Acrobatics. Conversely, if you're handing the reins over to a new pilot, and don't want to see your pride and joy thrown into the ground at speed, Loiter (Position Hold) will have your drone hanging in the air like a zen hummingbird.

As shown opposite, there are numerous levels of automation. Many pilots experience Loiter first (it is the default setting of a DJI Phantom, for example), and discovering that other pilots maintain altitude with constant manual adjustment of the throttle can come as a bit of a shock.

In addition to altering the way the craft responds, many systems now feature a number of fully automated modes, allowing the 'copter to follow a planned route or endlessly circle a spot, which is great for capturing video.

SMALL RACER
With no GPS unit, the choice of flight modes is more limited than on a system with that sensor fitted, but Altitude Hold will still make for an easier flight than Stabilize or Acrobatic.

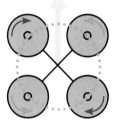

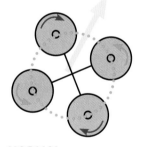

NORMAL
"Front" turns with the craft.

INTELLIGENT ORIENTATION CONTROL

IOC or "simple" mode is designed to help new pilots who often find it confusing to translate the controls to the current orientation of the craft (when a 'copter is facing toward them, all the controls are reversed). To achieve this, IOC uses the pilot's location as a fixed point, and all roll and pitch commands are referenced to that point.

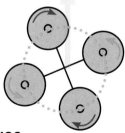

IOC
"Front" fixed away from pilot; craft rotates independently.

COMMON FLIGHT MODES

Name	Req	Behavior				
		Roll	Pitch	Yaw	Throttle	Description
Stabilize		Lean	Lean	Turn	Climb/Descend	The pilot controls the 'copter's pitch, roll, turn, and throttle, but the software will keep the 'copter from turning over.
Altitude Hold (Alt Hold)		Lean	Lean	Turn	Below 50% = descend; Above 50% = climb; Approx. 50% = maintain.	The 'copter will fly normally, but you can leave the throttle at the center and the 'copter will maintain it's altitude. Release roll or pitch and the 'copter will keep drifting like an air-hockey puck.
Loiter (Position Hold)	GPS	Lean	Lean	Turn	Below 50% = descend; Above 50% = climb; Approx. 50% = maintain.	When all controls are released and the throttle is left near 50% (stick in middle), the 'copter will maintain its exact position.
OF Loiter		Lean	Lean	Turn	Hover at 50%	As above, but uses optical flow instead of GPS.
RTL	GPS	-	-	-	-	Return to Launch. The 'copter will automatically fly up to a preset altitude, return to the take-off site, and land. Often used as a failsafe mode (if the RC fails, for example).
Auto	GPS					Follow a pre-determined route set up using software in advance.
Acro(batic)		Lean	Lean	Turn	Push (if craft is leaning, affects acceleration as well as altitude)	Similar to Stabilize, but hands full throttle control to the pilot allowing for acrobatics (and crashes).
Sport		Lean	Lean	Turn		In sport, the copter will not automatically stabilize; similar to the continual drift, as in Altitude Hold mode, but without maintaining altitude.
Drift (on APM)	GPS?	[right stick]	[right stick]	[auto]	Speed	Fly the drone as if it were a plane; it'll travel forward and you can move the "nose" up/down/left/right with a single stick.
Guided	GPS					The drone moves to specified points as the pilot clicks on a map, then hovers (as Alt Hold).
Circle	GPS					Orbits a fixed point, with the front of the 'copter always pointed toward it.
Land						Automatically brings the 'copter to the ground, then shuts off the motors and disarms.
Follow me	GPS					Follows a GPS unit worn on the "pilot," such as a cellphone.

THE MAIDEN FLIGHT

The first flight is the moment where pilot and machine become one; for the self-built 'copter it can also be a vindication of all your hard work. However, while we all want to be up in the air zooming around as quickly as possible, this flight—more than any other—needs caution.

I'VE ALREADY SAID "fools rush in" and at no point is that more true than the first time you fly any particular drone. I say that because however confident you are with one 'copter, the feel of those controls will not necessarily translate to another. It's vital to remember each 'copter has its own maiden flight—assuming one 'copter would be very much like another is a mistake your author has made and the results weren't pretty (there's a link to the video at tamesky.com).

Before you fly it's vital to check the instructions that came with the 'copter (or flight controller) as to the calibration procedure. Typically this will involve holding the 'copter and rotating it around all three axes in one manner or another (the "calibration dance") in order that the flight controller can be sure the gyroscopes know which way is up, which way is down, and so on. Failure to do this will result in an erratic (and short) flight with an unplanned end.

The magnetometer (compass) must also be calibrated. In some cases this can be done automatically by the software, in other cases you'll need to find out the correct settings for your geographical location and send them to the flight controller. That's because magnetic north isn't actually where polar, or "true north" is (in fact, it isn't fixed at all), so the flight controller needs to compensate.

It's also vital that the RC controller is correctly calibrated. This process involves rotating the sticks and releasing them so that the software can discover the range of movement and the center point of each

stick. Again, from excruciating personal experience, failure to do this will result in a 'copter flying in one direction (perhaps toward a tree...) while you're expecting it to stand still.

Calibration methods vary. For example calibrating a RC controller on a Arducopter/Pixhawk system requires you to connect the 'copter to a computer and then move the sticks, while on the DJI Phantom you connect the proprietary RC controller directly to your computer with a USB lead and use a specific app.

Once everything is calibrated, select the easiest flight mode available to you. If your 'copter has GPS then select Loiter or GPS hold, which are great for getting the feel of a 'copter as you can let go of the sticks and the 'copter will hold its position in space.

Otherwise Altitude Hold is the next best thing; this will use the air-pressure-sensing altimeter to keep the craft at the same height so you can get a feel for its motion using roll and pitch. From the moment of take off you need to be ready to make gentle movements with roll and pitch (in the standard Mode 2 that means the right stick) to hold position. If you have neither then "stabilize" is the way to go.

More than any other flight you'll need plenty of space around you at take off. You should stand at least 10 meters (30 feet) away from the 'copter, in case you need to correct any surprise lurches in your direction. Once you take off, the maiden flight is mostly about landing again!

BEFORE YOU FLY

Beyond all the usual precautions, make sure you've got a lot of space around you and the only people around are with you (or are "under your control" in the parlance of aviation authorities).

TAKE OFF

Take off firmly to give yourself some room to maneuver. Don't try and hover too close to the ground—it is far better to get a feel for flying when your 'copter is a few meters off the ground.

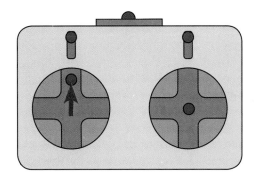

TEST RESPONSIVENESS

Move the craft around with gentle (and steadily less gentle) movements, keeping an eye on it to see if it performs as you expect.

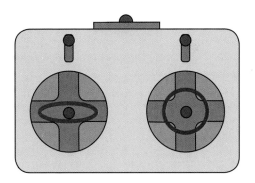

LAND

Once you are comfortable flying (and just before you start to push your luck!), land. Once you've landed, immediately disarm your 'copter. If your telemetry doesn't provide battery data, check the battery status to get an idea of how much longer you could have stayed in the air.

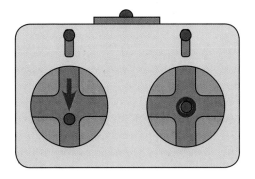

ROLL, PITCH, YAW, AND THROTTLE

03

ACE PILOT

Flying using the sticks (or their touchscreen equivalent) dusts off the lexicon of old-school fixed-wing aircraft, but you also need to adapt your mind to additional controls, such as those that enable vertical take off and hover.

PITCH

All other things being equal, pitching a multicopter will make it move forward or backward. On the commonly used Mode 2 configuration that means pushing the right stick forward to make the craft move forward. Personally, I would suggest taking off with the drone facing away from you, so pushing the stick forward will pitch the front down and the 'copter will fly away from your position.

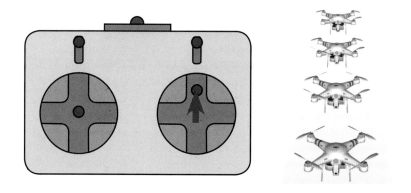

ROLL

Again, assuming the standard Mode 2 configuration, the other control on the right stick is roll, which makes the 'copter move left or right, while the front remains facing forward. The further the stick is from the center, the more the 'copter rolls and the faster it will travel. Many pilots find that getting used to the "right stick" controls (pitch and roll), while keeping their 'copter in a small area in front of them helps build confidence early on. This is especially true if the 'copter is in Altitude Hold or GPS Loiter mode. If video is important to you, rolling smoothly left and right is very useful for tracking moving subjects from the side, creating a form of aerial dolly shot.

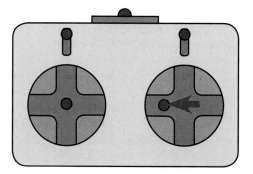

THROTTLE

The throttle determines how much lift the 'copter is creating: below a certain point the 'copter will descend (or stay on the ground if it's already there); above that point it will ascend. The ideal hover spot should be set to the stick's middle point, and on some ready-to-fly 'copters that default to Altitude Hold mode the sticks are even sprung to this point. In manual mode the throttle is much harder to master, as it doesn't default to a fixed altitude.

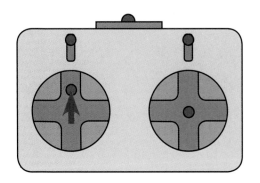

PITCH AND ROLL

Used in combination, the pitch and roll controls will always leave the "front" of the 'copter facing in the same direction, but it will slide around in the air in two axes (similar to a computer's mouse pointer). In the illustration below, the 'copter has been pitched at about 50% forward and 15% left, so it flies forward faster than it does left, but the flexibility is endless.

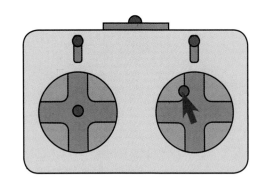

PITCH AND YAW

A different way to turn, which might look more elegant if you've got a forward facing camera, and will certainly be a better test of your piloting skill, is to combine forward movement from the pitch control with the rotation of yaw. Yaw turns the 'copter around its center, which it can do even when loitering in one spot, so only when combined with pitch does it create a linear flightpath. Once you've mastered combining pitch and roll, it's definitely time to add yaw to the mix.

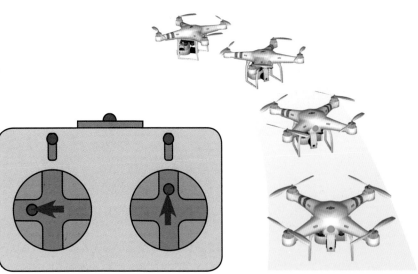

KEY SKILLS

Hovering and sliding around in straight lines is no fun, so once you've got to grips with the basic controls, it's time to move things up a notch. There are several skills that you should practice as often as possible to help you use the sticks more naturally and become a real master pilot.

THE REAL TRICK is the ability to use both sticks at once, and to develop an appreciation for the aircraft's orientation. This is strange at first, and there is a strong temptation to stick with Intelligent Orientation Control (IOC) or "simple" mode. However, doing so would be a lot like only learning to drive an automatic gearbox car in Europe, where stick-shift is far more common. Sure, you could drive if you found an automatic car, but you'd never be able to use most of the cars on the road and you'd certainly never get behind the wheel of a classic sports car.

Another tendency of many early pilots is to think only about one movement at a time, as if there were an invisible cubic grid guiding them. It's perhaps true that people with a photography background who see their 'copter as a flying camera find themselves flying as if they are moving an imaginary tripod around without much grace. Even for dedicated photographers this isn't the best way to fly; without developing more subtle control you'll never be able to shoot the swooping video that is such a big part of clients' requirements these days.

MANUAL ALTITUDE

The difference between Altitude Hold and manual throttle is pretty significant. It's less important for photographers than racers, but it's a good idea to practice trying to hold a fixed altitude without automatic assistance and as little input as possible from the right stick.

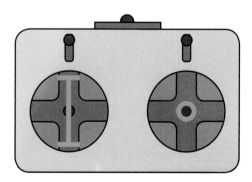

SIMPLE CIRCLES

Learning to fly in ever tighter and ever more accurate circles will help every aspect of your flying. Pitch forward with the right stick, then use yaw to turn the 'copter.

PASS-BY

Once you've got the hang of turning, it's vital to refine your grasp of flying in whatever direction the front is pointed. A good way to do that is make fly-bys of your position, turning tightly at either end; slow yourself with backward pitch. As you get better able to judge the direction of travel, you can make the straight flight longer and faster.

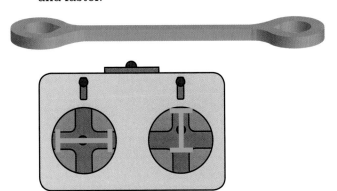

FIGURE OF EIGHT

Flying a figure of eight will require a combination of turning and passing-by skills. This will really help you focus on flying and perceiving your 'copter's position in 3D space. Try varying the speed and the size of the figure of eight for variety.

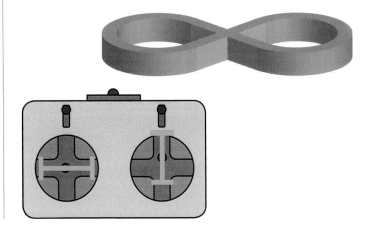

SIMULATORS

ACE PILOT

03

If you want to get used to the sticks without any risk, or you've been grounded but can't keep your imagination out of the air, a simulator might be the solution you're looking for.

THE BENEFITS OF a simulator are obvious, and the market has responded with plenty of choices. Drone manufacturer DJI has also started adding simulation modes into its latest 'copters—the DJI Inspire and DJI Phantom 3s both offer this functionality.

I've tried all these systems and I was surprised by how much they have to offer. I wouldn't say they accurately simulate the exact behavior of flight in any one specific 'copter, but as a way of getting a feel for how two sticks can translate into three-dimensional movement they're a great option. For beginners, that is the least natural part of learning to fly, so it's great to be able to practice safely.

One issue that you might encounter is with flight modes. Each app defines the key flight modes in its own way, and this is not necessarily the same way as your 'copter's flight controller. Consequently, if you get used to a non-stabilized simulator (the default mode on some sims) the real thing may surprise you.

NEXT CGM RC-HELI
This simulator can use a PS3 or PS4 joypad connected via USB, and the analog sticks translate perfectly to a real RC controller. If you have a console lying around, this is a great solution.

SIMULATORS

	Windows	Mac	Linux	iOS	Android
AeroSIM RC	✔				
Heli-X	✔	✔	✔		
Quadcopter FX				✔	✔

Visit http://tamesky.com/completeguide/ for links to these—and more—sims.

QUADCOPTER FX

Defaulting to Mode 2, you can use the on-screen thumb sticks to get a grip on roll, pith, yaw, and throttle. You can switch between an in-copter view and an on-ground view.

DJI SIMULATOR

Shown here on a Phantom 3 (but similar software is also included with the Inspire), DJI's simulator allows you to use the tablet or phone you use when flying and the same controller you fly with.

HELI-X

On top of simulating flight, Heli-X offers a series of challenges, such as flying between points or a reaction test. You need to plug a suitable control unit into your computer, though—not all of them have built-in USB ports.

MANUAL MODE

What's often called "manual," or "rate" mode still makes just as much use of the flight controller as the others, it's just that the way it chooses to level (or not) the 'copter makes it more exciting to fly. Manual works on virtually any system—it's the preferred mode of many racers.

PART OF WHAT'S great about manual mode is that it's the lowest common denominator. You can spend a lot of money on camera ships—for which stability and ease of use are vital—but you can easily add a racer to your fleet for a much more modest sum.

Full manual doesn't just mean that you need to maintain the altitude yourself, it also means you have to counter any pitch or roll that you apply in one direction yourself by pushing the stick in the other direction (as opposed to stabilize mode, which levels the 'copter when the sticks are at the mid-point).

FLIPS

Although it's certainly not recommended for massive multicopter's hoisting expensive movie-making equipment, lighter and more sporty craft can easily be pitched or rolled a full 360 degrees, which can be pretty spectacular. This won't be quite as controlled as the automatic flip function of the AR.Drone, so a quick hard boost on the throttle before you flip should give you the altitude you need to make the turn without powering into the ground. Be confident before you try this.

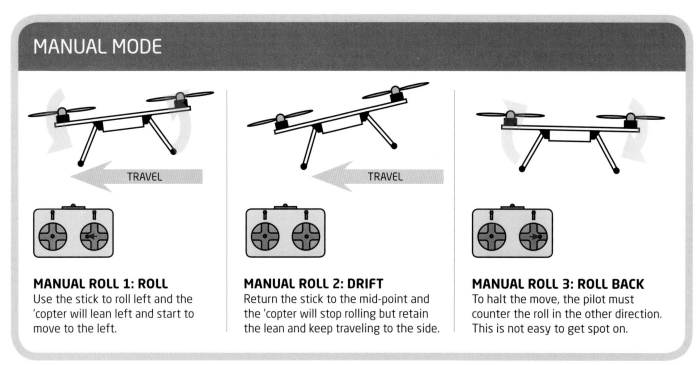

MANUAL MODE

MANUAL ROLL 1: ROLL
Use the stick to roll left and the 'copter will lean left and start to move to the left.

MANUAL ROLL 2: DRIFT
Return the stick to the mid-point and the 'copter will stop rolling but retain the lean and keep traveling to the side.

MANUAL ROLL 3: ROLL BACK
To halt the move, the pilot must counter the roll in the other direction. This is not easy to get spot on.

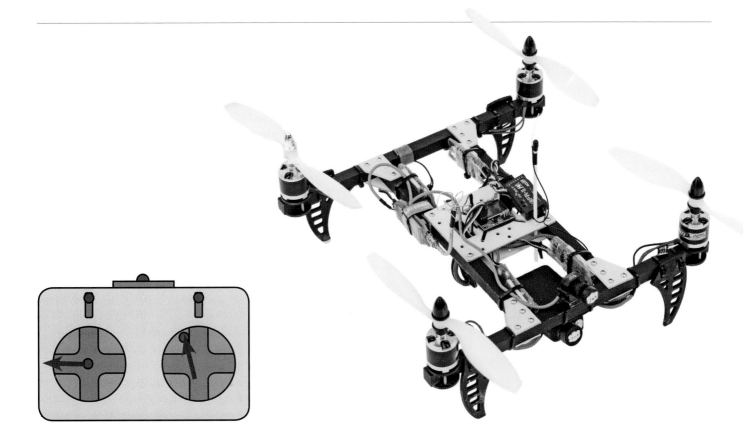

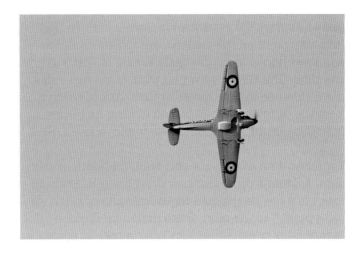

LEANING INTO TURNS

Learning to fly in ever tighter and ever more accurate circles will help every aspect of your flying. Pitch forward with the right stick for speed, use the yaw on the left for the turn, but don't be afraid of pushing the right stick the same way as the left one to roll into the turn. Practice this with circles and figures of eight.

The further you lean into a turn, the more of your 'copter's energy is devoted to sideways motion, rather than lift, so you need to be confident that you have the power to maintain altitude—you may need to push the throttle up to compensate for a loss of altitude, or even ensure your altitude is high enough before a fast charge forward (which will cost altitude).

RC FLYING
Manual mode has much in common with traditional RC flying, although rolling a traditional plane (or performing a loop-the-loop) doesn't involve pointing four propellers toward the ground!

BUILT FOR MANUAL
A 'copter built for manual is one that has cheap, replaceable parts. Although this bright quadcopter by Jack of Brighton looks to be a quite complicated construction at first glance, it is made of cheap wooden pieces (even the legs), which are held together with cable ties and standard screws. Breakages are frequent in manual mode, so replacement needs to be easy—ideally something you can do on-site and for a low cost.

GPS AUTOPILOT

Paranoid journalists often peddle stories about drones as if they all had (evil) minds of their own, when the reality is an RC is usually in line of sight. Autonomous mode gives you the opportunity to bring (some) of those journalistic fears to bear by pre-programming your flight path.

GPS SATELLITE
One of the 24 satellites forming the GPS "constellation." Russia and Europe have also deployed GLONASS and Galileo.

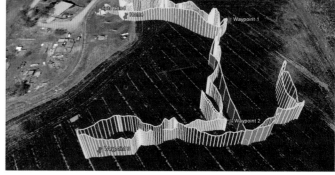

FLIGHTPATH
A 3D diagram showing a drone's flight-path, recorded in flight and then viewed using Google Earth.

THE GPS SYSTEM is made up of a network of satellites with incredibly accurate clocks on board, which broadcast the time. Their near-Earth orbits mean they are constantly moving around the earth (unlike geo-stationary TV satellites) in such a pattern that, from any point on the surface, around 10 satellites will have line of sight (for radio waves at least) at any one time. The GPS receiver on your drone measures the difference in time those signals take to reach it from the different satellites, calculating its position from this—including altitude—to within a few meters.

Together with a compass and other on-board sensor data, a drone's processor has everything it needs to execute a pre-programed flight from one position, or "waypoint," to the next. Even in normal flight, a GPS sensor adds possibilities like loiter mode, in which

the drone will hold its position when you let go of the controls, making it great for hovering.

Thanks to the incredibly detailed mapping available online, planning your flight can be as simple as point and click. However, GPS isn't perfect. For a start, it is only accurate to within a few meters, so your waypoints are more like invisible spheres (see page 79). There are also more physical concerns; when planning your waypoints you need to make sure the route between them avoids buildings and trees, and also pay close attention to your range (likely to be determined by the battery).

SETTING WAYPOINTS

Setting your waypoints for an autonomous flight—without user input—can be as simple as clicking at points on a map using an iPad then defining the altitude you want to hit at that point. Here, the DJI software is shown. While it is an elegant solution, it not only requires a DJI flight controller, but also needs the expensive addition of a datalink with Bluetooth OS.

AR.DRONE FREEFLIGHT

If you have the optional GPS module you can set waypoints and track your flight with the standard app...

QGROUNDCONTROL

...however, free apps are also available that can be used to set a fully autonomous journey.

03

ACE PILOT

STEP 1: LOAD THE MAPS

In this example I'm using the DJI Phantom 2 Vision+ in which, once you've enabled it in the settings, you can swipe left to reveal a waypoint planning tool.

You might think that you can connect your iPhone to your copter, receive the GPS location from the copter and get going, but having the link to the Phantom via Wi-Fi confuses the iPhone into assuming it'll receive internet via Wi-Fi rather than the cellular connection. Instead I found I had to start up as usual so the location could be found, then disable the WiFi and wait for the map tiles to load in the DJI software, before logging back into the Phantom's WiFi signal.

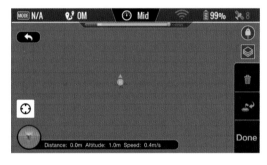 

STEP 2: PLOT WAYPOINTS

Click on the map at the points you'd like to fly to. The on-the-ground distance from take of is shown so you can check that it's realistic – it should be within your line of sight so you can immediately take over with the controller if there is a problem, and take into account local rules about airspace, people and property.

STEP 3: ALTITUDES

Click on each waypoint individually to set its altitude. If you're passing over an obstacle – say a tree – make use that the waypoint before it is set to the tree's height so as to be completely sure that it isn't clipped in the assent even if the next waypoint beyond the tree is already set above the obstacle's height.

3D WAYPOINTS

Software typically sets virtual spheres around waypoints; as soon as the copter feels that it's inside one of these spheres it'll move onto toward the next one (the solid line here representing the path the copter would likely take, the dotted the planned route). Although this can lead to some inaccuracy, setting the sphere size (waypoint diameter or waypoint radius) too low might mean the copter hunts in the sky near the first waypoint without ever reaching it.

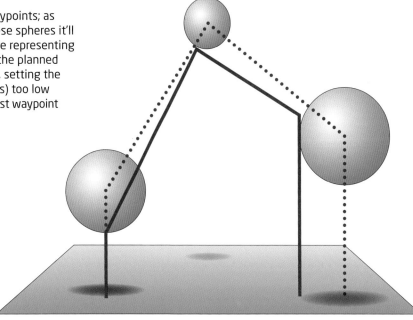

STEP 4: CHECK AND FLY

Once you're finished check all the waypoints. Other platforms also allow you to carry out instructions at waypoints, for example to rotate or to circle the location, both useful for photography, so make sure each waypoint has the correct command associated with it. Then stand well back and make sure the route is safe and clear and you have the controller ready to take over if necessary and click 'Go'.

3DR TOWER APP

Available for android devices which are capable of powered USB is the radio serial link for very sophisticated waypoint planning. A word of warning: A Samsung Galaxy Tab S won't provide power to the USB port but a much cheaper Google Nexus does. Check DIYDrones for a list of compatible tablets.

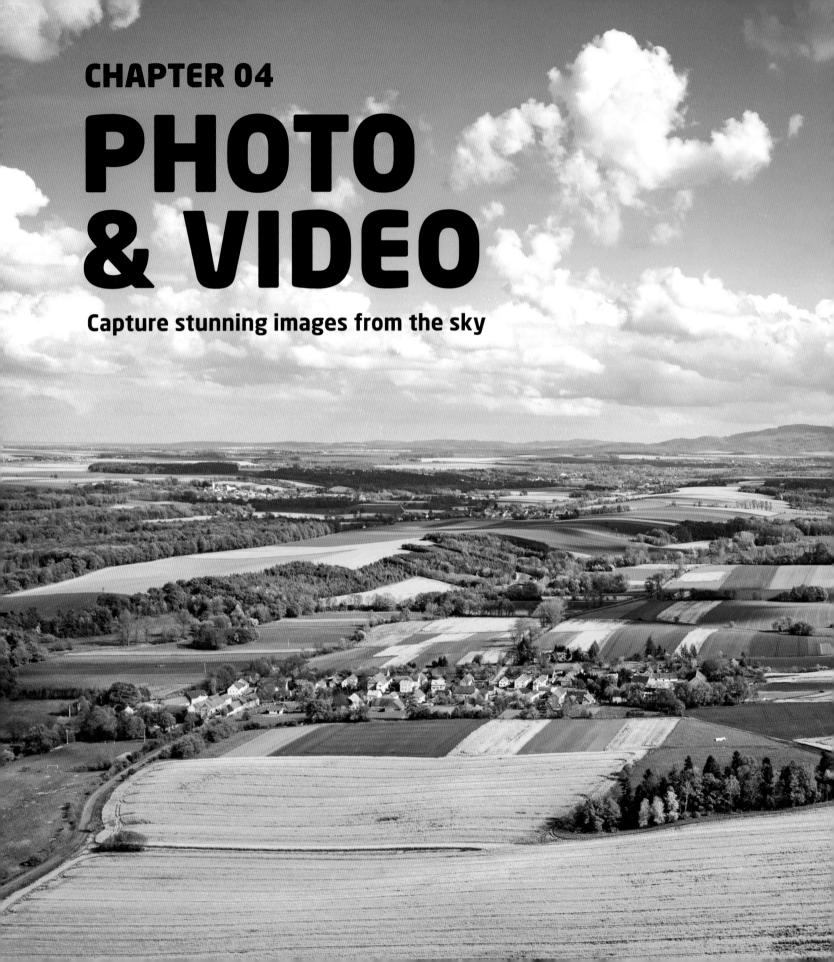

CHAPTER 04
PHOTO & VIDEO
Capture stunning images from the sky

CAMERA SPECIFICATION

Even if you think you know about cameras, flight's emphasis on weight puts a whole new range of equipment in front of you. How should you decide between megapixels, video resolutions, field of view, and bit-rates? What are the trade-offs and what's essential?

FOR MANY PHOTOGRAPHERS the big change between choosing a digital camera and choosing a camera for a 'copter (or a multicopter with a built in camera) is that the emphasis switches from still image specifications like megapixels to factors affecting video quality. Professional photographers also need to take note that the ability to switch lenses is, at the time of writing, unknown in 'copters save those heavy lifters that offer support for existing SLR and interchangeable lens cameras

SENSOR MEGAPIXELS

Experienced photographers hear the word megapixels so often that it begins to lose meaning—arguably the word is used because manufacturers found it exaggerates the effect of their slight improvements to camera sensor quality. Nonetheless it's an important measure of a camera's likely quality of still images (assuming they are all printed at the same size).

VIDEO RESOLUTION

Video quality stems from both the number of pixels—essentially the same as megapixels but usually defined by key standards like 1080 and 4K—and from the number of new frames a second. Computer enthusiasts will find traditional broadcast, or 'standard definition' difficult to get their heads around since the only true 'pixel' measurement is vertical, or 'number of lines.' The actual picture is transmitted as an analogue signal which means that when NTSC or PAL (the names for the American and European standards respectively) is converted for use on a computer the pixels are not necessarily square.

The other slightly tricky thing to comprehend about NTSC and PAL is that they are interlaced, meaning that only every other line of the picture is updated with each frame, a system that worked well with old cathode ray tube TVs. Digital video on computers and eventually HDTV for the home brought with it the possibility of refreshing the whole frame at once, just as in the cinema, and this is known as 'progressive' scan. Within the world of drones you'll likely encounter NTSC/PAL only in the very analogue world of FPV goggles.

HOVE
This seaside view was originally captured with the DJI Phantom Vision 2+ 14MP camera.

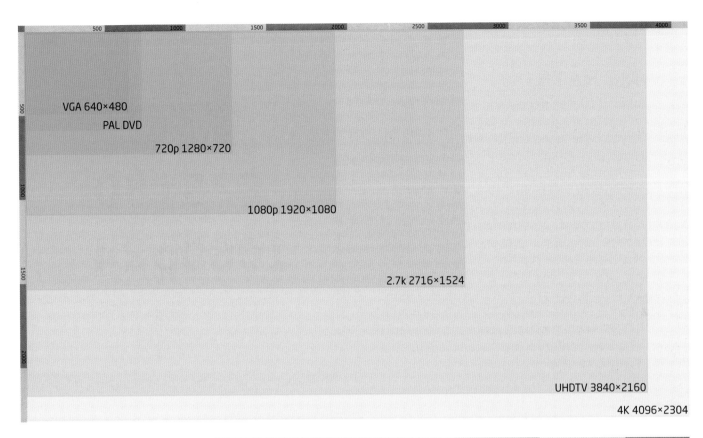

VGA 640×480

PAL DVD

720p 1280×720

1080p 1920×1080

2.7k 2716×1524

UHDTV 3840×2160

4K 4096×2304

VIDEO RESOLUTIONS

There are several different standards for video. Traditional US TV is similar to VGA, but European PAL was slightly higher. The 'HDTV' wave of the mid 2010s had two key standards; 720 and 1080 (measures of the vertical pixel resolution) and now UHD 4K & Cinema 4K (both often called 4K) are emerging standards.

STILL RESOLUTIONS

Still images are usually captured in 4:3 (old TV) or 3:2 (traditional photo) proportions; this graphic gives an idea of how much more information can be found in progressively higher megapixel counts.

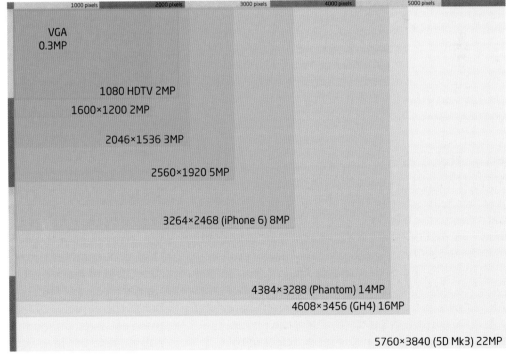

VGA
0.3MP

1080 HDTV 2MP

1600×1200 2MP

2046×1536 3MP

2560×1920 5MP

3264×2468 (iPhone 6) 8MP

4384×3288 (Phantom) 14MP

4608×3456 (GH4) 16MP

5760×3840 (5D Mk3) 22MP

CAMERA SPECIFICATION

04

PHOTO & VIDEO

FRAME RATE

The number of frames per second has a dramatic effect not just on the quality of the video but the flexibility you have when editing with it. You'll often want to slow the footage down to create cool effects, and if the original has a fast frame rate you can do this and keep your video silky smooth.

In terms of perception, 24 frames/second (progressive) has long been enough for cinema, though on some level viewers are aware that the effect created is 'cinematic', which is why Peter Jackson kicked off such a big debate when he tried High Frame Rate (HFR) with the Hobbit movies. TV's rates of 50 (PAL) and 60 (NTSC) seem perfectly adequate, so a rate of 120fps which will still run without any judder on a 60fps TV when slowed.

Many cameras will offer you a trade off between frame rate and resolution as their processor chip can only encode a certain amount of data at once, and they'll use this to advertise a higher resolution than will realistically be useable. The first GoPro cameras to offer 4K only did so at 15fps, well below the 24fps of cinema, the lowest useful speed. I'd suggest that you consider the effective maximum video resolution of a camera to be the one it can offer at your preferred editing rate.

BIT RATE

The one thing that almost no one mentions is bit-rate, which is a measure of the amount of data recorded per second. It's a little like the difference between long play and ordinary VHS, except the lower quality manifests itself in digital artifacts—less detail in dark areas, softer edges, things lingering on screen for a few moments and digital blocking). The Parrot AR.Drone 2 is a good example; although theoretically a 720p30 camera offered HD video, the bit rate is very low so the quality is severely affected by

---

READING VIDEO SPEC

625 i 50
1080p 24

Vertical lines   Progressive/ interlace   Frames/Sec

STANDARD	FRAMES/SEC
Traditional Cinema	24
HFR Cinema	48
PAL	50i
NTSC	60i

artifacts. External cameras like the GoPro offer a choice of bit rates (qualities) in their settings to save card space.

FIELD OF VIEW

Photographers are used to adjustable zoom lenses, but the weight of the extra optics required means that most cameras drone users will fly with have a fixed field of view, including those built into popular 'copters. A large field of view allows you to see more but at the expense of fish-eye distortion. GoPros offer three settings, although the widest is the sensor's natural field of view, at which it'll operate best (in

IMAGE SENSOR

This is a life-size diagram of different image sensors; from the center outwards: DJI Inspire / GoPro Hero Black, then Micro 4/3rds camera, then 35mm 'full frame'.

WIDE-ANGLE V RECTILINEAR

A picture shot with a GoPro set to wide angle, and the same view from a rectilinear lens.

VIEW FIELD

Different fields of view affect the image captured.

118°

GoPro Wide / Phantom 2 Vision+

94°

DJI Inspire 1/Phantom 3 and GoPro medium

64°

GoPro Narrow

40°

Often taken by photographers as broadly equivalent to human eyesight (excluding peripheral vision).

terms of low-light sensitivity and still image resolution). An alternative, 'rectilinear', is still distorted, but will feature flat horizontal and verticals.

SENSOR SIZE

The size of the image sensor (the chip which does the job film used to) is important too. Typically described by its longest measurement, the diagonal (like TVs), bigger is generally better in terms of image quality because it allows for larger individual pixels on the chip (called 'photosites'). The bigger they are, the less noise in the image. GoPros and DJI cameras both use sensors of 1/2.3" in size, bigger than the iPhone 6's 1/3", though new DJI sensors have slightly lower resolution, a trade off which provides bigger photosites.

EXPOSURE TRIANGLE

Camera technology boils down to the three ways a camera controls the light; shutter speed, aperture and sensitivity (ISO). From vastly expensive digital cinema cameras through SLRs, built in cameras, right down to basic FPV cameras, these three controls are always present.

WHOLE BOOKS HAVE been written on the subject of exposure. The best of them always make two things easy to understand: the fundamental connection between these functions, and the fact that you're probably used to the camera handling them for you (and that might not be the best idea).

Imagine that the exposure triangle is real and miraculously balancing horizontally on a central point. If you press one corner down, then the other two will rise, or if you push one down and keep another at the same point then the third corner will rise further. That's the principle at play; three values with a single sum.

Irritatingly photographers measure the three values separately: shutter speed, measured in fractions of a second, aperture, measured as f-stops, and light sensitivity, measured in ISO standardized numbers. The combined setting is known as the Exposure Value (EV), and in many cases you might want to ask your camera to adjust the EV up or down (to lighten or darken the image). EV is generally measured in 'stops', the chart indicates full stops on each of the three key measurements.

The bar chart indicates how the process works; the dotted line indicates the 'perfect' exposure (not too light, not too dark), and the bars are made up of aperture, shutter & ISO settings. The last two examples are 'overexposed' (too light) and 'underexposed' (too dark).

Of course three different settings bring three different sets of trade offs. Setting a wide aperture (low f-stop number) lets a lot of light in, so you can use a high shutter speed, but wide apertures reduce the focal length, which means that not everything will be in focus. This is hugely significant with high end cameras, and the artistic look it can achieve is part of the reason photographers spend so much on wide-aperture

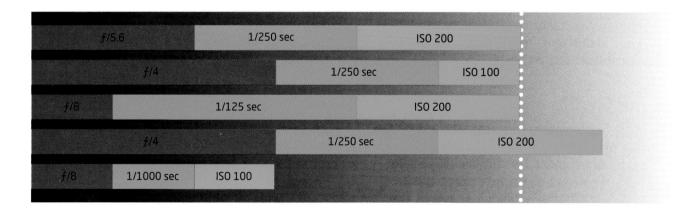

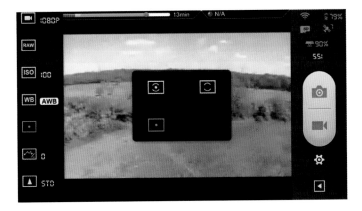

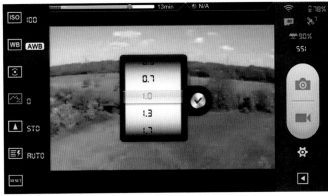

METERING

Your camera will attempt to set the exposure automatically, but you can tell the camera whether to bias toward the center (center-weighted), take an unbiased view from the whole frame, or only take measurement from the center spot.

PLUS OR MINUS EV

To lighten the image as a whole, you can ask the camera to increase the Exposure Value (EV).

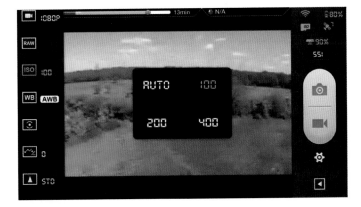

MANUAL ISO

Setting your camera's ISO manually will take that choice away from the camera's auto mode; instead it'll need to resort to whatever control it has over shutter and aperture. Bigger photosites make higher ISO settings work better (see page 87).

lenses, but with small-sensor cameras the effect is much less pronounced.

Higher shutter speeds lead to crisper, clearer images as they freeze motion, but the faster the shutter speed the more light you need. Increasing the sensitivity of the image sensor to light with a higher ISO setting seems the natural solution, but the higher the sensitivity, the greater the noise (a grainy pattern of image degradation).

EXPOSURE VALUES

This picture shoes the effect of exposure values from -4 to the left to +4 to the right.

EQUIVALENT STEPS ON EXPOSURE SCALES

F/STOP STEPS	SHUTTER STEPS	ISO STEPS
f/1.4	1	100
f/2	1/2	200
f/2.8	1/4	400
f/4	1/8	800
f/5.6	1/16	1600
f/8	1/30	3200
f/11	1/60	6400
f/16	1/125	128000

CASE-STUDY: AMOS CHAPPLE

Amos Chapple has flown above Spain, Russia, Iran, India and more besides with the aim of presenting great architectural achievements in a way never seen before. "This is the way buildings are designed to be seen, this is their best side" he says. He certainly brings the models to life.

"SO FAR THERE hasn't been a lost of stills photography and I was amazed by that" said Amos Chapple of the revolution in the creative industry that has been wrought by the arrival of drones (and perhaps, we might speculate, the marketing of 'copters to people like him—people who are defined by the creative role rather than tech enthusiasm).

Amos is sensitive the need to find a free space without people beneath the 'copter. "I've not crashed it, though there were some difficulties with monkeys in India". Additionally there are some situations you can't use it, such as slums where it would excite much interest. He says "you need three or four or five flights" to identify a good shot and capture it.

In terms of the future his outlook is slightly pessimistic about future regulation, "which is why I've been doing so much so quickly." "The window is very much closing," he feels, because "creativity and rules don't mix." That's a shame because he already refers in the past tense to "a very brief golden age of drone photography."

TAJ MAHAL
"I was advised of the rules [for aerial photography around the Taj Mahal] and then I had to leave very quickly" he says, intriguingly.

RED SQUARE
This was a commissioned shot from a Russian publishing company, but they made it clear I was on my own if I got caught.

SACRÉ-COEUR
Paris' Sacré-Cœur glowing in a hazy sunrise. Weather is everything for aerial work, the images which have made an impact have always been the pictures where there is something unusual happening with the light. This was a dawn that at first looked like it was going to be overcast— I remember being within an inch of going back to bed.

© amoschapplephoto.com

CAMERAS

PHOTO & VIDEO

The drone world has a few favorite camera models; the all-conquering GoPro is amongst them but it's not the only one. Indeed the price of the GoPro and its flat-on proportions don't lend it to smaller racing quads, where the compact Mobius dominates.

MAX VIDEO RESOLUTION: 4K
MAX FPS: 120 (@1080p)
Field of view: 133.6°
41 x 59 x 30mm
88g (83g for Hero 3 series)

MAX VIDEO RESOLUTION: 720P
MAX FPS: 30 (@720p)
Field of view: varies with model
50 x 32 x 16mm / inches
38g (1 ¼ oz)

GOPRO

The original 'action camera,' the GoPro Hero series is popular with broadcasters and action enthusiasts because it was the first camera designed to make capturing sports footage easy while staying close to broadcast standard. It didn't hurt the company's image that the founder outwardly epitomized the surf lifestyle, talking up the money he raised selling beads (and mentioning rather less the huge investment from his parents).

There are regular refreshes to the GoPro, but the Hero 3 and Hero 4 have most support in the gimbal and follow-me drone market; any changes to the form factor (shape & positioning of buttons and ports) that post-date this book might pose problems, so make sure you're selecting a compatible GoPro.

808

This bargain keychain video camera, which can be found at online stores for about the price of a Big Mac meal, is a favorite for fliers expecting to crash! It's rugged and cheap, which is where its appeal lies, but its certainly not for the faint of heart; typically found from retailers who don't provide a lot in the way of instruction, especially not in a language you can read.

There are numerous versions of this product – the 720p version was #16 – which means you need to check the specs carefully when ordering from your hobby supplier. A downloadable Windows app can be used to alter camera settings.

MAX VIDEO RESOLUTION: 1080P
MAX FPS: 60 (@720P)
Field of View : 116 at 1080p, 63° at 720p60
61 x 35 x 18mm / 2½ x 1½ x ¾ inches
38g (1 ¼ oz)

MAX VIDEO RESOLUTION: 4K
MAX FPS: 96 (@720P)
Field of View :Depends on lens
133 x 93 x 84mm / 5.2 x 3.6 x 3.3 inches
560g (19.75 oz) without lens

MOBIUS

A successor to the still-popular 808 series, the Mobius embraces the "action camera" status with faster frame rates, improved picture quality and Full HD. The camera is sold with a clip-case that features a standard tripod mount. Additionally a number of mounting attachments and housings are available for cycling, watersports and so on.

The metal 'buttons' are actually a heat sink, allowing the camera to disperse the heat the processor generates while encoding video. A useful feature for pilots is that the video can be recorded either way up, so the mounting isn't too problematic. As with the 808, Windows (and Android) apps access settings.

LUMIX GH

The Panasonic Lumix GH3 and GH4 cameras are popular because they use the well established Micro Four Thirds lens system, an interchangeable lens standard agreed between several manufacturers. This means you will have no trouble finding a good lightweight lens for aerial videography.

The Lumix offers a high bitrate, a generous 100mpbs (mega bits per second), but the data rate is the same for 96fps as 24, so the quality is lower at high frame rates. It also offers 48fps, exactly double the classic movie frame rate making editing together with pro footage easy.

GIMBALS

04

PHOTO & VIDEO

In order to maintain flight, multicopters need to lean this way and that; if they're using GPS to maintain position they'll be constantly being battered by the wind, hardly ideal conditions for photography. Motorized gimbals are the solution; countering the drone's motion for stable photography and videography.

CAMERA GIMBALS ARE generally accepted as the only way to secure good quality video & photography from the skies (though admittedly Parrot might take some issue with that). Gimbals are all about balance; in order to keep the weight to a minimum, the load on the motors shouldn't be too high, which means that, all things being equal, the camera should simply hang level beneath the 'copter. Beneath, by the way, is usually the optimal choice since the camera is a heavy component and stability is enhanced by keeping the center of gravity lower down.

The main difference in the market is between two and three axis gimbals; a two-axis gimbal will hold a camera steady and allow it to be panned up or down (sometimes known as the roll and pitch planes), a three-axis gimbal adds rotation (yaw). It's also worth noting that gimbals and cameras are built as single components in many modern drones (for example DJI Phantom 3 and Vision series) while they remain distinctly optional on others, like the 3DR Solo.

The motors respond to high-speed measurement of the conditions to keep the camera level, and in addition it might be possible to use a radio channel to control the pitch and even yaw. For example the DJI Phantom 2 and 3 feature 3-axis stabilization, but you can only tilt the camera up and down; yaw is used to soften sharp turns and for stabilization only.

If you're building your own craft, it's important to make sure that your gimbal will be able to receive a control signal from your RC; typically you'll be able to allocate a channel from your RC's flight controller to the gimbal's computer and set it up so that it pans the camera up or down using the same motor that dampens the drone's motion.

Although ready-to-fly kits will include both the gimbal and the control technology, self-builders will be able to chose separately control hardware (like the AlexMos) and gimbal components, allowing a wider choice of cameras.

ELECTRONICS
This is the opened housing for one of the three motors on a DJI Phantom's built-in gimbal. It features motor control circuitry and devices to measure and correct for the copter's angle.

DIGITAL GIMBAL
The Parrot Beebop uses an interesting alternative to the gimbal; it has a high quality wide angle lens built in but it automatically adjusts the crop that is recorded to simulate a gimbal without the extra weight; the effect is surprisingly good.

DJI ZENMUSE

DJI offer a range of gimbals called Zenmuse which are designed for specific cameras; their popularity is such that they influence which cameras are commonly used on multicopters.

FRONT-MOUNTED GIMBAL

The Team Black Sheep (TBS) Discovery Pro copter places the gimbal at the front, ideal for smooth footage from exciting high-speed flights. Great for racers, though admittedly gimbals are more delicate than the drone's speed-fans they're used to!

REMOVABLE GIMBAL

The Yuneec Q500 Typhoon comes with a novel feature; a removable gimbal and smartphone grip so the expensive gimbal can be used on the ground too (at time of writing there is some evidence DJI will do something similar soon).

STILLS COMPOSITION

That drones can produce some stunning aerial photography won't come as a surprise to readers of this book. What you need now is to find a way of standing out. Here are some pointers.

LOOK DOWN
In this shot, ©Rus Turner directed the camera down to eliminate the horizon; no matter, we see the sky reflected in the water and gain a more interesting view into the bargain.

LOOK DOWN
Jase Wickman captured this stone circle.

TECHNICAL VIRTUE
Rather than try to correct the lens distortion, why not emphasize the distortion with a digital filter as Paul Miller has.

ROLLERCOSTER

Alessandro Magnoni has got some great shots of this rollercoaster at Kemah Boardwalk, Texas, but the larger shot is definitely the winner. Providing context is unnecessary, so the straight-down view is more interesting, but the final choice has to be the shot that rewards careful examination: if you look closely you can see the carts on the track.

SHIP BOW

Flying over the water guarantees that should something go wong there's no chance whatsoever of recovering your 'copter, but capturing the bow of a ship like this, the RMS Queen Mary 2 visiting Australia. Note the use of the classic rule of thirds to set the horizon and look down on the deck. © Lynh Phan

VIDEO COMPOSITION

04 Although the more you fly the more you'll discover your own ability to create stunning swooping aerial movies, when you're getting started there are a number of classic shots you'll want to practice.

AERIAL DOLLY

Although a dolly track can be any shape, the classic one is a straight line following along a moving object. With a multicopter, though, recreating this kind of shot is easy (so long as you're in a mode that holds altitude for you). Simply rotate to face the subject and use roll rather than pitch to glide left or right. Experienced pilots might find that altitude hold, rather than GPS hold, will provide the smoother dolly, though be warned—the 'copter won't stop when you release the stick.

CIRCLE

By orbiting a fixed spot and keeping it in the center of the frame it will apparently move very little in the frame while the surroundings move quickly. The narrower the field of view and the further away the background, the faster it will appear to movement in the frame. This kind of effect is used to great effect in Michael Bay's *The Island*, among many more. To achieve it, rotate the copter (and camera) to face the subject, then circle with roll rather than pitch, keeping the subject in place with the yaw.

Pic ©Matin Underhay.

CRANE CAM

A great staple of the TV studio production (in fact it's impossible to picture a live broadcast of a stage event without cranes over the crowd and the performers), the crane helps producers achieve swooping but level camera. Add a multicopter and suddenly the crane is infinite; you can pan across a seafront while rising over the buildings and providing a constantly altering perspective in three axes at once; a very engaging shot, whatever your subject, but you'll need to use roll, yaw & throttle and perhaps camera pitch all at once, which requires practice!

THE DRONIE

OK, you nearly got away without my mentioning it, but it's become a classic shot since Patrick Stewart's "dronie", taken at Cannes 2014, in which the 'copter starts close to the subject so at first sight might just be a normal steadicam, then suddenly lurches up and back to reveal the surroundings as quickly as possible. The inevitable @Dronie account on twitter now has thousands of followers despite somewhat sporadic maintenance.

A word of caution; make sure you fly the correct direction—backward when the camera is front-facing as it is on most copters—as well as up, and keep the camera facing down at about 45°.

DUAL CONTROL

Dual control is a professional feature that has been the mainstay of high-end video production for some time, separating the control of the aircraft and the camera equipment makes it easier for production teams to get the shots they need first time without risking safety.

FLYING AN AIRCRAFT while keeping your eyes on the framing of your shot is difficult no matter how good the flight controller's stabilization. You need to have your eyes in two different places at once while operating the 'copter (four axes of movement) and the camera (one or two more). It's not impossible, as the deluge of stunning videos online demonstrate, but the addition of the second controller streamlines the process.

Not only does this arrangement play to the strengths of the respective operators, but it allows for more sophisticated camera functions than mere orientation to be controlled remotely. One such example is the follow-focus. This is not a function of automatic focus as the name implies, but the movie-maker's term for an arrangement of gears which allows smooth shifts of focus. With a remotely operated follow-focus one could achieve a beautiful cinematic focus pull from a foreground object to the background, something the automatic or fixed-focus lenses on most drones doesn't allow for.

Dual operators get better the more they work with each other. The pilot and camera operator will have to start by describing their intentions for each shot and agreeing whether it'll work, aiming to keep the client—the film producer—happy by keeping the number of takes to a minimum and set up time

DJI INSPIRE DUAL CONTROLS
The DJI Inspire, seen elsewhere in this book, can accept single or dual controls; there is only one camera, though, so the pilot must use line of sight and the map view.

PILOT, CAMERAMAN & SPOTTER
The HeliVideoPros (HeliVideoPros.com) have worked with Fox, Sony and more, and their team consists of dual operators and a 'spotter' to keep an eye on any potential risks and warn the pilot.

DUAL CONTROLS
Sarah and David Oneil
(of That Drone Show fame –
see Chapter 6) operating
a DJI Inspire 1, In this shot
David is piloting while Sarah
is directing the camera.

DUAL CONTROLS
A pilot and camera operator
working together side by side.

between them low too. Mercifully (and for obvious reasons) aerial video very rarely involves the need to record sound, so crew chatter presents no problems, but eventually crews will develop a language of their own, familiar with types of shot and with a shared history to refer back to. This sort of partnership will be valuable to a filmmaker which is why you'll often see crews advertising their many years experience. Admittedly those in the know may spot that their time working together predates the arrival of this kind of 'copter, but a working partnership is always a big, albeit unquantifiable, asset.

The DJI Inspire 1's dual control mode, and especially the ability to add the second controller at a later date, has taken what was once a function specified by professionals in their build-to-order craft and made it more wallet-friendly. Although this is still far from a cheap aircraft, it is considerably more accessible than most truly professional 'copters. Until that point by far the cheapest way to create a dual control mode was to self build, not impossible but requiring an additional radio receiver to be added on board, all adding complexity.

You'll note that most dual control craft are bigger beasts; this isn't really to accommodate the extra technology, more that the kind of situation where dual control is needed means weighty professional cameras will be involved.

YUNEEC TORNADO H920
The professional level Yuneec Tornado is designed for single and dual operator flight and to carry the filmmaker's favorite micro-four thirds camera: the Panasonic Lumix GH4.

PROGRAMMED CAMERA

By taking advantage of the drone's sensors and "brain" you can get more controlled (and more elegant) shots than you might be able to achieve in the moment with the sticks. You can get video as good as working with a second operator without needing one.

Only a couple of years ago the distinction between enthusiast and professional flight controllers was really whether the software was all about the drone or whether the filmmaker's needs were covered too. That was distinction was very clearly delineated by cost. Even now most of the pre-programmable flight-paths are simple waypoint-to-waypoint ones in which the camera will turn with the front of the drone to the next waypoint. There's no real technical reason for that, most drones can face any way equally well after all, it's just how the software has developed.

That, however, is starting to change. The 3DR Solo's software marked a big shift from what came before it, deliberately designed to include shots like their "Cable Cam", in which the 'copter flies along a designated path allowing you the freedom to turn the 'copter to face your subject. The result is just like a cable camera over a stadium, creating a very cinematic shot without too much thought on the pilot's part. Circle and other modes are supported too.

Another form of remote camera control becoming increasingly popular is the Follow Me mode. Various PR teams are falling over themselves to claim invention of this feature for their particular brand (credit should probably go to 3DR). In any case the essence is this; the 'copter follows the position of the operator on the ground via a device they're wearing, an android phone in the case of the Iris+ (summer 2014), or a remote control bracelet in the case of the AirDog (spring '15), or back to the app for the Hexo+

CABLE MODE
The tablet screen mounted to the 3DR Solo's controller shows a green virtual cable on which the 'copter will remain fixed. It can fly from one end to the other and be rotated to point the camera in any direction, great for big-stadium video over pee-wee hockey games (assuming everyone consents to the drone's presence).

(fall '15). Despite the obvious enthusiasm for this technology it's not without its problems—these copters are not generally equipped with object avoidance technology (save for optional optical flow on the Solo, and that only points downward). It remains to be seen just how good they are at keeping your GoPro (the camera of choice for each of these copters) away from objects or, in the case of action sports, the ground or water. Use with caution (or a generous budget for repairs!)

AIRDOG FOLLOWING DRONE

The AirDog will follow you wherever you go, as loyal as a flying dog for as long as the battery lasts. Controls on the wrist bracelet allow you to take off, land, and change the pattern.

3DR SOLO

The American company's answer to the DJI Phantom, the Solo takes programmed camera options to the next level, with everything the filmmaker could wish for.

GHOST DRONE

The Ghost was launched with app-only control, though an RC unit is an optional extra. As such all flight directions are essentially programmed, and follow-me is an option.

IMAGE CORRECTION

PHOTO & VIDEO

When you're capturing video a fish-eye lens makes a good amount of sense; the wide field of view ensures you won't miss anything. When it comes to printing or sharing your picture or video, you're more likely to want to tell a story with it, which means using software to correct those problems.

RAW FILES

Digital cameras actually discard a great deal of information that their sensor records when they create a JPEG image, and that image will likely be created using the camera's best guess at settings. Even if you've set manual settings, a great deal more detail can be recorded if you save the image as a RAW file rather than a standard JPEG (billions of possible shades instead of millions).

The difference might sound academic, but when you start pushing colors, shades and contrast with the tools in an image editing program you'll quickly find that a JPEG doesn't look so good. A RAW editing program will usually allow you to tweak the Exposure by at least the equivalent of a couple of stops as well as re-thinking the White Balance.

Once you've finished editing a Raw file the changes are preserved in a 'sidecar' file (just the changes, not the data), but saving out to a JPEG is the best way to share your work with the rest of the world.

STILL CORRECTION

Photoshop and Lightroom recognize popular cameras, including the wide-angled DJI Phantom 2 Vision+, and can correct for lens distortion at the click of a button.

FINAL CUT PRO

There are a number of plugins available to assist with the correction of curvy horizons. One of the most popular is available from Alex4D.com and is yours for the price of a donation towards its development.

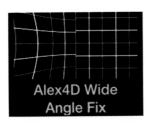

Once installed it creates an extra effect you can apply to any clip in Final Cut Pro X. It includes a full set of presets for GoPros, or if you're correcting the DJI Phantom 2 Vision+ You can choose the GoPro 127° preset which is pretty close. Alternatively if you can't match your lens precisely select 'Custom' in the drop-down and you can adjust the correction yourself. You can even change the colors of the on screen guide grid if needs be.

PREMIER PRO

Adobe's flagship video editor provides a useful cross-platform editing tool and in the 2014 CC version they added support for GoPro and DJI Phantom lens correction without leaving the program.

Once you've got your clip in the timeline, simply open the Effects pane, and in the Presets folder you'll find Lens Distortion Removal folders. Open your camera's brand and model and then drag the correct file to your clip (the filed of view of the camera, and hence the amount of correction required, varies depending on your choice of image resolution amongst other features. If you find you select the wrong one, you'll need to undo or remove it before applying another, or both will be applied at once.

CHAPTER 05
BUILDING

**Build your own super-cheap 'copter,
the "Wooden Wonder."**

TAMESKY 1: THE WOODEN WONDER

Until now, we've concentrated on why and how drones work, and we've seen a lot of ready-to-fly drones pictured. However, many people build their own 'copter, starting out with nothing more than some basic building materials. If you're interested in having a go yourself, then why not try our "wooden wonder?"

FOR THIS PROJECT, I got together with Jack Nash (known online as "Brighton Till I Fly") to talk about the most crucial skills you need when building a new drone, as well as ways that the cost can be kept down. In Chapter 2, almost every possible component was discussed, but for this 'copter, the "TameSky 1" we've pared things right back to the essentials.

At this point, it's worth appreciating that Jack is both a skilled engineer and a very able FPV pilot, and is committed to driving costs down by using his expertise and ingenuity. I, on the other hand, am able to see the advantages of spending a little more money to bypass some of the "more involved" moments. With this in mind, there are a few points in this build where Jack and I would have slightly different recommendations based on cost versus effort—both options are given.

We're building this 'copter out of wood, so if you chose to build a multicopter with a carbon airframe, some of the details in this chapter won't apply. However, the essentials on power distribution and flight controller hook-up are the same, regardless of the materials used. Note also that many RC parts are measured (and sold) in millimeters, so metric measurements will be used throughout.

DOWNLOAD PLANS
To download the designs for this 'copter, plus a list of parts, visit:
www.tamesky.com/building-drone

JACK'S DESK
Your work environment is entirely up to you, as long as you know where to find everything you need. If we are totally honest, that was not always the case during this build.

TOOLS

You will need:
- Hacksaw/hand saw
- Soldering iron and solder
- Screwdriver and 3mm screws
- Drill with 3mm bit and 17mm spade drill bit
- Spray adhesive
- Holesaw (2-inch/51mm)
- Wood file
- Craft knife

PARTS LIST

For an updated version of this list, please check the book's webpage at www.tamesky.com
- 4 motors (we used 1100kv Prop Drive Series 26–28 from HobbyKing) and accessory packs (accessories are often a separate product)
- Props
- 4 ESCs Hobby King 20A ESC 3A UBEC (you could select slightly more expensive ESCs that have been pre-flashed with SimonK firmware, but Jack prefers to do this himself)
- KK-Mini flight controller board from HobbyKing
- 1.2m 10x10mm stripwood
- Birch plywood sheet
- Battery connector
- Radio controller & receiver
- Battery grips
- Hook-and-loop battery straps
- Adhesive hook-and-loop strip
- Double-sided adhesive mounting foam
- Heatshrink for cables
- Zip ties (cable ties)

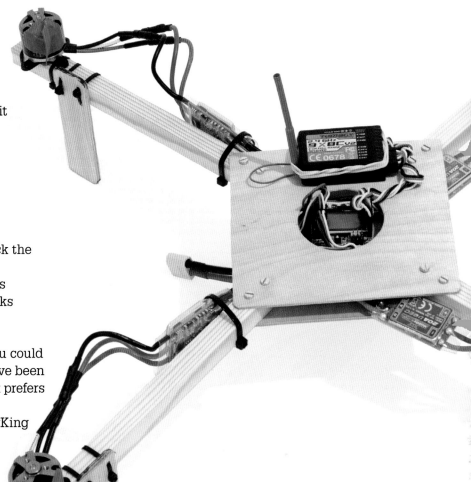

THE TAMESKY 1 BUILD (PART 1)

The first stage of building the 'copter is surprisingly basic—it's simply cutting and crafting wood. If you look at this section and see other ways you might approach them, that's great; the point of the "wooden wonder" is to show just how basic a multicopter can be—innovation is encouraged!

1: PRINT AND CUT

This design features a central hub, and templates have been supplied to make cutting these easier. Start by printing out the templates at www.tamesky.com/building-drone, or photocopy them from pages 143–144. Use spray adhesive to attach the paper templates to your plywood and cut around the sides with a handsaw (thin plywood can also be cut using a sturdy craft knife).

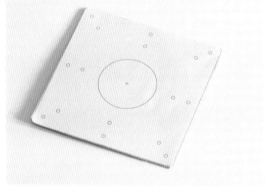

At this point you should also cut the arms to length (we chose 220mm) and stick the hole-cut guides onto one side of each.

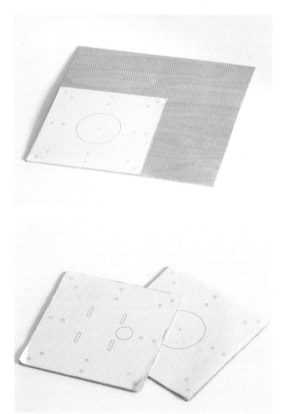

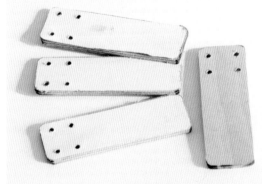

2: DRILL

The next step is to drill the pre-marked holes on the boards. Use the correct drill bit for your screws (here 3mm), as you want a good firm connection. Jack uses a pillar drill mount with a Dremel multitool, but any household drill should be adequate for the task. Take care to drill the holes in the correct position, though—this is crucial for keeping the frame symmetrical.

In addition to the holes indicated on the lower plate, you also need to drill a row of small holes that you can smooth out with a file— these will be the holes for the battery straps.

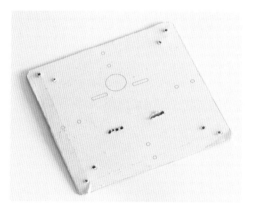
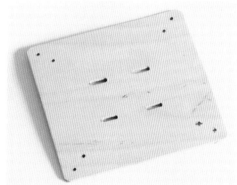
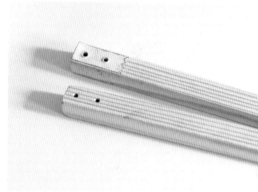

3: LARGE HOLE

The lower panel also needs a larger hole cut in it. For this, use a 17mm spade drill bit, which works just like a regular drill bit. Make sure that you place (and ideally clamp) your panel over a large piece of spare wood, as the drill will drive a long way into the surface beneath before the wide spade area cuts the circle through which the battery connector cable will run. It is also likely to cause a little bit of splintering around the edges of the hole, which will need to be sanded or filed down.

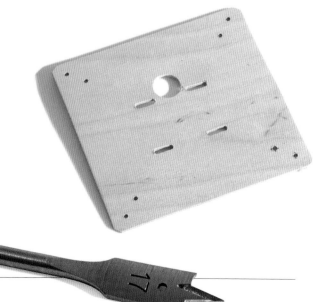

05

BUILDING

4: HOLESAW

Moving to the upper plate, you need to make a large hole in the center, through which you can place (and reach) the flight controller. This is so you can protect the relatively delicate control board from any impacts. Use a holesaw of a suitable size (2in./5cm was adequate here) to drill the hole, but be aware that the rotation of the saw will cause the paper template to be ripped away—it will help if you drill the smaller holes first, and then secure the template around the edges with masking tape.

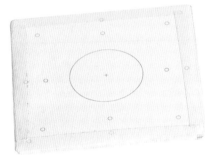

5: ASSEMBLE THE FRAME

Just eight screws/bolts with nuts are required to hold the frame together. It doesn't hurt to make sure everything fits together, so we pushed the 3mm screws through the holes and loosely prepared the final frame. There's no point tightening anything at this stage, though, as you will be taking it apart again to fit the electronics.

6: BATTERY STRAPS

This is an opportune moment to fit the battery straps. The slots in the TameSky 1's lower hub piece are designed to take two hook-and-loop battery straps of the type pictured.

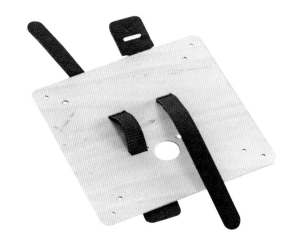

7: MOTOR MOUNTS

Brushless motors spin in such a manner that the entire cylinder rotates, so they can only be mounted from below. That means that each motor needs to be fitted to a motor mount. This is a small X-shaped piece, with holes for short screws to go up into the motor's barrel, and holes to screw it into the model. These mounts form part of the motor's accessory pack.

FLASHING ESCS

"Flashing" ESCs (updating their software) is very much a personal choice, but it's important that all your ESCs are running the same firmware. It is also helpful if the firmware is the multicopter-friendly SimonK. You can buy ESCs with the firmware already set to SimonK software, which is a perfectly valid solution.

AS I MENTIONED at the start of this chapter, Jack and I have different views on different things and this is one of those areas. Personally, I've always opted for pre-flashed ESCs, and for a one-off project, that's probably the cheaper solution. However, if you plan on making multiple 'copters (or are part of a club that shares resources) a firmware flashing tool will save you money in the long-run (a flashing tool costs about the same as one ESC).

The other advantage of flashing your own ESCs is that you can take advantage of improved software, rather than making do with the software installed by the manufacturer (and who knows how long ESCs have been in warehouses).

SimonK's firmware works by increasing the refresh speed of the ESC, possibly from as low as 8kHz to over 400kHz. This means that your motor will get instruction from the flight controller not a "mere" 8,000 times a second but a much more useful 400,000. In flight, this provides a massive improvement in stability.

Note that SimonK firmware shouldn't be used with anything other than a multi-rotor craft because the throttle control to the motor is made instantaneous. This removes the softening (averaging) function that would protect components from harm on an RC helicopter.

Links to Mac and Windows software can be found at: tamesky.com/book

1: REMOVE SLEEVE

ESCs might look like enclosed units, but their protective seal is usually just heat-shrunk on and can be gently cut away with a scalpel or craft knife. Do this with each ESC, taking care not to push the blade any further in than necessary, as it could easily damage the delicate components.

SLEEVELESS ESC
The ESC with its sleeve removed.

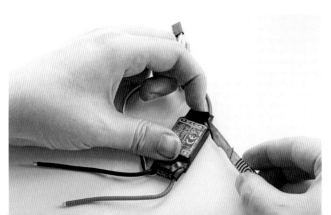

2: SETUP SOFTWARE

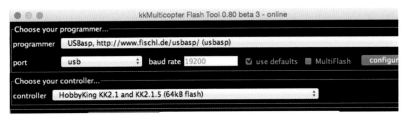

Connect your flashing tool to your computer's USB port and launch your software. There are a number of apps available for download, some more technical than others (check the book's website for links). We're choosing one of the easiest to use, which allows you to select the programmer (the flashing device you're using) and the type of controller chip that you will be flashing. The same app also allows flashing of the flight controller, so be sure to choose both *USBasp* as your device and *atmega 8-based brushless ESC* as your target controller.

3: LOCATE FIRMWARE

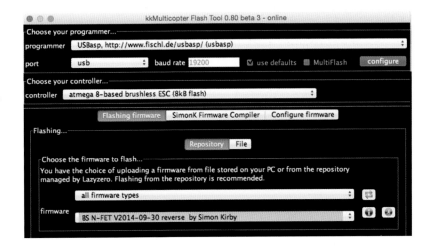

With the target identified, you will be shown a repository of possible firmware that you can download and install directly. We use BS N-FET V2014-09-30 by Simon Kirby (the legendary SimonK). There are two versions—forward and reverse—but the motors we're using allow us to switch directions simply by swapping a power lead. This means we can flash all four with the same version if we choose.

4: APPLY TOOL

Locate the chip on the ESC and press down on it with the flashing tool. The red dot on the tool should match the marked corner of the chip. Now, click the green icon to write the new firmware to the chip.

5: RESEAL ESC

Once you've successfully updated the firmware, place a large piece of heatshrink around the ESC and heat it using a hairdryer to safely reseal the components. It doesn't hurt to add the original sticker over so you remember what component you're using, too.

CIRCUIT DIAGRAM

In the next part of the build you'll get involved with the electronics, but it's important to have a good idea of why you're connecting it the way you are. Although the number of motors (and partner ESCs) might vary, and you can add accessories, the basic wiring of a multicopter remains the same.

THE KEY TO the diagram shown opposite is the power distribution. You might decide to use separate batteries to power the motors and other components, but assuming you use the right ESCs, there really is no need. That's because, as well as suppling power to the motor, the ESC also supplies power back to the flight controller along the red and black wires of the three-wire ribbon; the white carries the signal instructing the ESC how fast to turn.

Similarly, power is passed onto the radio receiver in this way. On the KK-Mini board that we've used it's not even possible to plug in the normal 3-pin connector leads to each channel; instead you must use just one for each of the radio channels for roll (ailerons), pitch

(elevator), yaw (rudder), and throttle—this is the signal wire. The board does feature a battery monitor and alarm, though, which requires a single lead run from the positive to the sensor pin.

One thing to mention; swapping any two of the three leads between the ESC and the motor causes the motor to spin on the opposite direction. The ability to change this is vital as you'll need to ensure that all the motors turn in the correct direction. This is why I like to use bullet connectors for these joins (the alternative is to make permanent connections and flash the ESC with 'reverse' firmware to reverse the motor). This seems needlessly fiddly though for the sake of a few easy-to-add connectors.

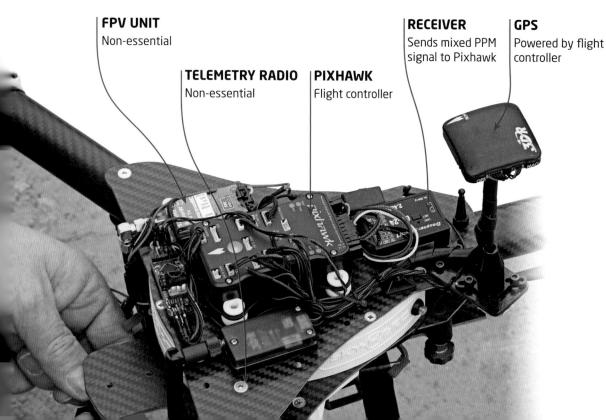

FPV UNIT
Non-essential

TELEMETRY RADIO
Non-essential

PIXHAWK
Flight controller

RECEIVER
Sends mixed PPM signal to Pixhawk

GPS
Powered by flight controller

SAME ON THE INSIDE
Although there are many additional devices, the hub of this Y6 reveals that the top of the Pixhawk flight controller has a series of servo cables headed to ESCs just like any other 'copter.

TAMESKY 1 CIRCUIT DIAGRAM

The Tamesky 1 has the most simple power distribution of any quadcopter—the only "luxuries" are the buzzer for audible feedback from the KK flight controller and the battery monitor lead. Other setups may add additional units, perhaps even additional voltages via a converter from the power distribution board, but this is the core functionality.

The number of cables between the radio receiver and the flight controller will depend on whether an analog signal (PWM) is used—which is the case here—or a joined digital PPM signal is used. The latter needs a single cable, plus the power connection, as discussed in Transmitters on p46).

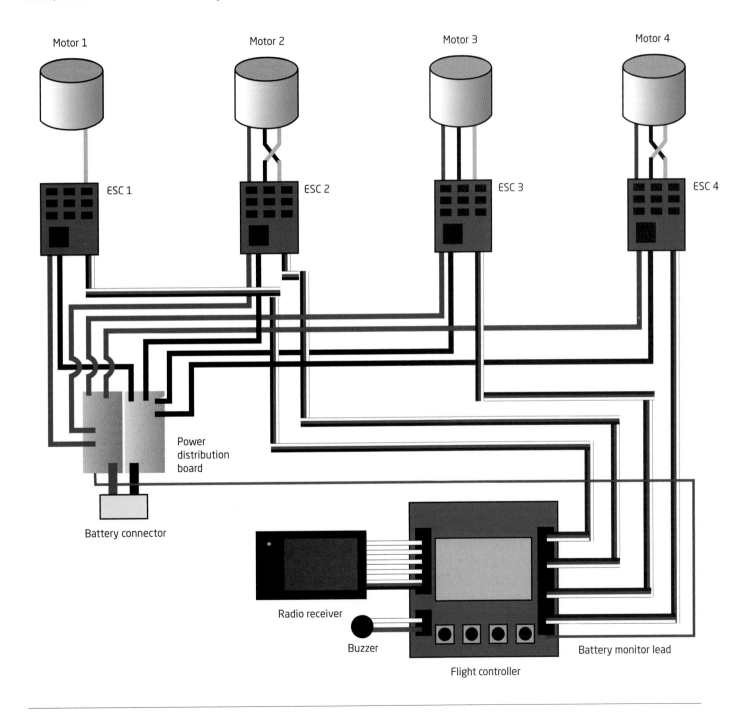

Motor 1 Motor 2 Motor 3 Motor 4

ESC 1 ESC 2 ESC 3 ESC 4

Power distribution board

Battery connector

Radio receiver

Buzzer

Flight controller

Battery monitor lead

THE TAMESKY 1 BUILD (PART 2)

Having taken a couple of pages to look at the technical side of things, you can get on with implementing that into your "wooden wonder's" frame, continuing from where you left off on page 113. The first stage is to sort out the wiring, before getting the flight controller up and running.

8: BULLET CONNECTORS

At this point it's important to fit all the bullet connectors that are needed between the ESCs and motors, You might also choose to place them between the power distribution board and the ESCs, but that's an extra step that we're going to side-step for the TameSky 1. The popular 3.5mm connector size is ideal for the kind of current we'll be dealing with.

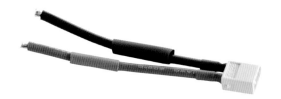

9: BATTERY CONNECTOR

Locate the battery connector and add generously sized heat shrink wraps to both leads. This will be the starting point of your power distribution "tree." You'll also need all of the ESCs and a piece of bare wire. The type of connector you choose will depend on the battery type—the JST-XH is popular for quads.

PCB POWER DISTRIBUTION

In this build we're going to use the simplest form of power distribution possible: connecting the wires together with solder and a gripping wire, then sealing the join with heat shrink. That is not the only method, though. You can buy dedicated power distribution boards (one for a 'copter like this will cost about the same as a Happy Meal), or you can create your own using a piece of copper-coated PCB (pictured, based on an idea from Bruce Simpson at rcmodelreviews.com). You could even make one each for positive and negative using clean, conductive washers.

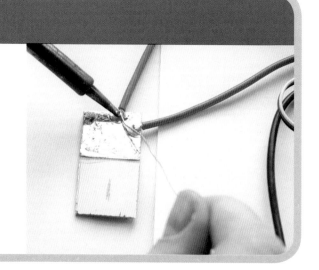

BULLET CONNECTORS

One thing you'll encounter a lot with multicopter builds is bullet connectors. These are a safe and easy way to connect components such as motors and ESCs, without creating a permanent link. In fact, they're especially useful between a motor and ESC, as they make it easy to swap the wires if you need to reverse the motor direction (you won't know the direction until you power it up, so this is handy).

Soldered bullets

Soldering your bullet connectors provides a firm connection, and is the method we recommend. Soldered bullet connectors also tend to be shorter, making for a more flexible joint (the length of the male and female connectors, when joined, forms a straight, inflexible section that can be fiddly on some airframe designs).

To solder a bullet connector, drill a hole (or several holes) in an offcut of wood that is just large enough to accommodate the connector. Push the connector into the hole and heat the connector with your soldering iron. When it is hot enough, apply the solder so it forms a pool of molten solder in the reservoir. Making sure that both the tip of the wire and the bullet connector are heated, dip the wire into the pool of solder and wait for it to cool. Note that blowing on it to make it cool faster can lead to a less secure connection.

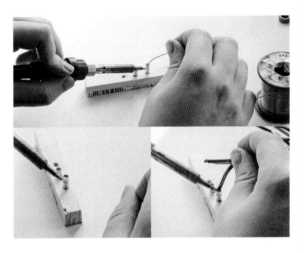

Crimped connectors

An alternative to a soldered connection is a crimped connector. These connectors simply clip over the cable and exposed wire and are clamped shut with a crimping tool (similar to pliers) to form a tight seal. This can then be soldered if you want a doubly secure connection.

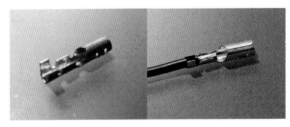

Heat shrink

To protect your connectors, apply some heat shrink. Make sure you cut pieces that are slightly longer than the connector and that it lines up exactly with the end of the connector when you apply heat. You don't want any part exposed to avoid the risk of short circuits.

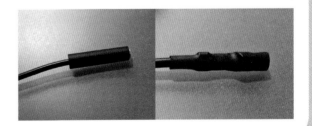

THE TAMESKY 1 BUILD (PART 2)

BUILDING

10: POWER DISTRIBUTION

The aim here is to connect each of the four positive (red) wires from the ESCs and each of the four negative (black) wires from the ESCs to the relevant battery connector wire. Note that the motor side of the ESC has three wires (which should already have bullet connectors for the motors), while the battery side has two wires (red and black)—make sure you work with the right wires.

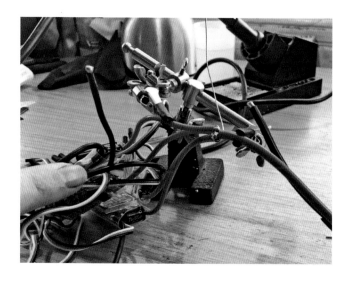

A word of warning—this is a bit fiddly, and is an area where you can choose between cost and convenience. Personally, I would opt to buy a dedicated power distribution board, whereas the only investment needed for Jack's approach (outlined here) is time.

Start by applying solder to the ends of all of the wires individually (positive and negative from both the ESCs and battery cable). Then, using a crocodile clip clamp stand (if you have one), or some weights to keep the wires in place, bundle together the wires of the same color— we're starting with the positive (red) wires. The four wires from the ESCs should be at one side, with the wire from the battery connector at the other, as shown above.

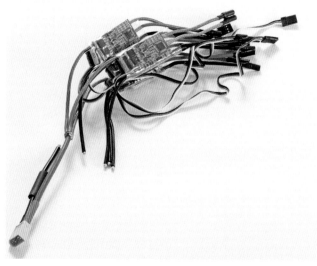

Wrap a couple of turns of bare conductive wire around the join, and then apply more solder to create a bond. Ideally, the connection should allow the power to travel from wire to wire, rather than relying on the solder to carry the current.

11: SEAL THE JOIN

Once you've got a strong connection, test it by applying gentle force to make sure that if the wires are under a little strain it doesn't fail. Assuming all is OK, pull the heat shrink tubing you put in place in step 9 up over from the battery lead and apply heat to make the seal.

Repeat steps 10 and 11 with the other set of wires.

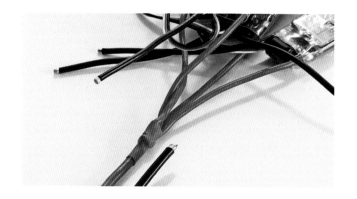

12: FIT THE POWER TREE

Once you've connected the positive and negative leads, you're ready to attach the tree of ESCs you've created to the multicopter's frame. With the arms attached to the lower hub board, pull the battery connector through the large hole in the board, so the JST connector passes out of the bottom of the hub. You can then attach the ESCs to the arms using zip ties (near, but outside the hub, where they will stay cool).

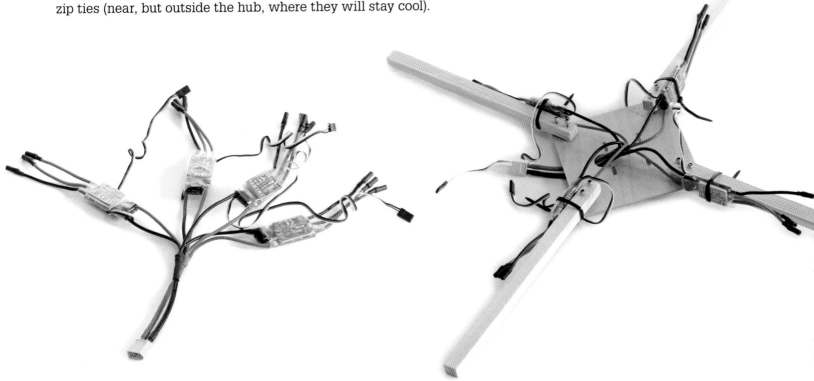

13: ATTACH THE TOP PLATE

Pull the thin control leads from the ESCs up through the large hole in the upper hub plate, then tighten the two parts of the drone together.

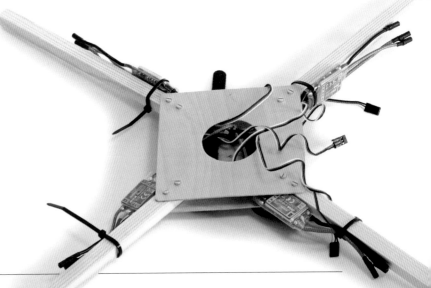

THE TAMESKY 1 BUILD (PART 2)

05

BUILDING

14: BATTERY

Attach a strip of adhesive hook-and-loop fastener to your battery and a corresponding one to the bottom plate so the battery will not slide away from the central position (this helps maintain the center of gravity).

15: CUSHION

Using a few layers of double-sided adhesive mounting foam, cut into short strips, build a padded platform for the flight controller to sit on. This will absorb vibrations and protect the motion sensors.

16: FLIGHT CONTROLLER

Place the flight controller through the large hole at the center of the hub. A small arrow on this board points toward what will be the "front" of the 'copter, so for an X-frame design like this you should make sure that the board is parallel with the outside of the hub. This mini board also features an external buzzer, which you can see dangling from the white connection point. Make sure you connect this, as it provides useful feedback.

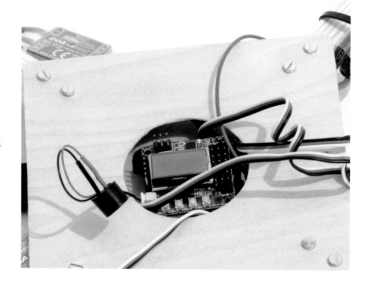

17: CONNECT THE ESCS

With the flight controller in place, it's time to start connecting things up. The ESCs not only provide power to the motors, but to the flight controller board as well, so before you can use the flight controller you need to plug the ESCs in. If you've already connected the motors, make sure they don't have props on, as you'll need to spin the motors a few times as you set everything up.

The ESCs connectors must be plugged into the correct ports to work. From top to bottom of the flight controller board are 3-pins for each motor, up to a maximum of eight—we'll only be using the first four for this quadcopter. Your motors are numbered from the top left (motor one) in a clockwise direction, as shown below.

The pin closest to the screen on this board is the signal wire (this is usually white).

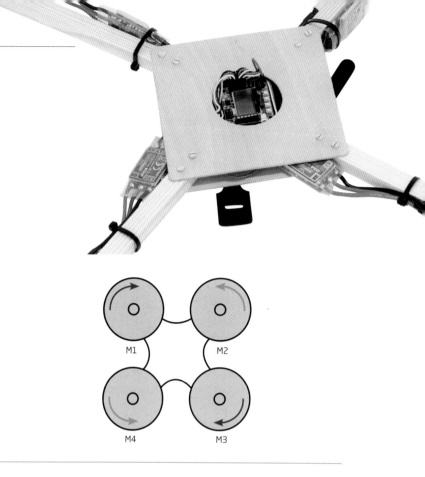

18: CONNECT THE RECEIVER

In this example, our receiver outputs a signal on each channel, traditional PWM, which is the more complicated method when it comes to wiring (a PCM system can push up to eight channels down a single signal wire).

Start by connecting the pins at the left side of the receiver from the bottom up. For the first three, you can use a single servo lead that plugs into the three pins of Channel 1 on the receiver. For the remainder you only need to connect to the signal pin on the receiver, which is why a three-wide cable has been used "sideways" in the picture below.

These connections mean that the flight controller board can receive power from the ESCs (when you connect a battery to your drone) and share that power with the receiver via the + and - pins at the left. This is vital for the next step.

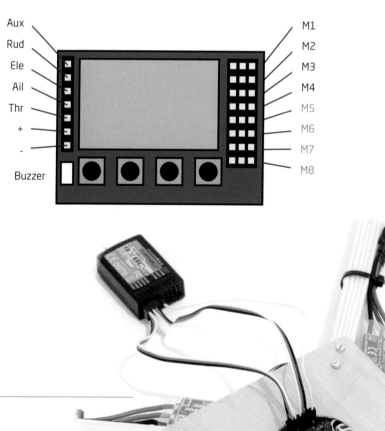

19: BINDING

The next step is to bind your radio receiver to your transmitter, for which the battery will need to be connected. Check the instructions supplied with your radio, but this is usually achieved by fitting a bind loop to a specific set of pins on the receiver while providing power to it. This loop tells the receiver to look for a signal from the transmitter—with my system, I press a button on the back of the transmitter and the blinking red light goes to a steady red to confirm the signal. (Note: I've disconnected all but one servo lead from step 18 to show that's all that's needed to

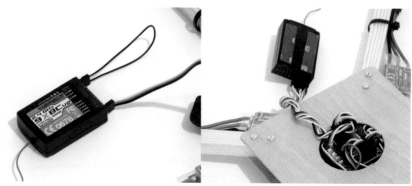

provide power). Once this is done, remove the bind loop, twist all of the control wires together and use some double-sided adhesive mounting foam to stick the receiver to the top of the hub.

20: MOTORS

Now is a good time to attach the motors. These need to be equidistant from the hub, close to the end of each leg. Grip each end of the motor firmly with a zip tie—this may not seem solid, but it's easy to repair after a crash.

21: LEGS

Having seen the screen light up with a battery connected at step 19, you can feel confident that you don't need to dismantle the hub, so now is a good time to add the legs. These are designed to absorb a solid landing and to be easily replaceable, which is why we're attaching them with zip ties rather than screws. Push zip ties through diagonally opposed holes in the corners of the legs, as shown above.

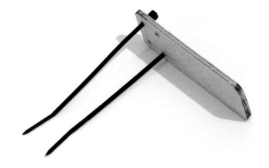

22: LEGS

Attach the legs just inside the motor position and secure the cable ties. Pull them very tight, then clip the ends off.

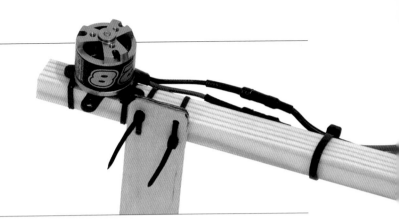

23: TAPING THE MOTORS

In the next step we'll be setting up the software, but first we need to determine the direction the motors are turning (so we can correct if necessary). This is surprisingly difficult to do, because of the motor's speed and symmetry, but attaching a piece tape to the rotating external cylinders can make it a lot easier.

You can see the correct direction of rotation in step 25—if you find you need to reverse the direction a motor spins in, simply swap any two of the three leads connecting the motor to the ESC.

24: OPEN MENU

Configuring the KK flight controller is fairly easy, as the four-button menu interface is surprisingly intuitive for something that looks so technical. However, it does throw some jargon at you. When you plug in a battery, you should see a screen like the one shown. Start by pressing the button beneath the word MENU. (Don't worry if 'ERROR' appears where 'SAFE' does here)

```
SAFE
Self-level is OFF
Error: no Yaw input.
Battery: 0.0 V
 Roll Angle:
Pitch Angle:        MENU
```

25: SET 'COPTER TYPE

As the eight motor connectors suggest, the KK board can handle a number of different configurations. For this design, choose *Load Motor Layout* from the menu and then select *Quadrocopter X*. The screen will then show the motor layout (which you can revisit via *Show Motor Layout*).

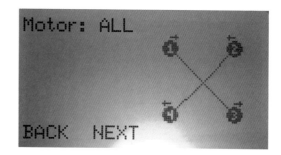

```
Motor: ALL



BACK   NEXT
```

26: GEAR CHANNEL

From the *Mode Settings* option make sure that *Self-Level* is set to *AUX* rather than *Stick*. This means that you can use the 5th channel on your transmitter/receiver to switch self leveling on or off.

```
Self-Level    :
Link Roll Pitch: Yes
Auto Disarm  : Yes
Receiver     : Std
Channel Map  : No
Lost Model Alarm: Yes
BACK PREV NEXT CHANGE
```

27: RECEIVER TEST

It's vital to make sure that all the channels are linked to the correct inputs on the board so the control sticks will do what you expect. Make sure your transmitter is powered up, select *Receiver Test* from the menu, and move the sticks, making sure that when you push each stick the correct response appears on screen.

```
Aileron   :35    Right
Elevator  :-1
Throttle  :0     Idle
Rudder    :-2
Auxiliary :-97   Off
Arm Test  :Safe Zone
BACK
```

If necessary, swap the channel signal wires from the receiver to the KK board so the correct stick affects the correct setting. After that, make sure that pushing the aileron right makes the word *Right* appear on screen. If not, you'll need to reverse the stick's function using your radio transmitter's menus.

28: SUBTRIM

You also need to ensure that the channels center at zero and reach 100 at the furthest point of stick travel. Using your transmitter, push the sticks to their furthest positions and adjust the sub trim according to the menus on the transmitter. Unfortunately, this varies from one transmitter to another and there isn't space to cover them all here (or anticipate future models).

```
Aileron    :0
Elevator   :0
Throttle   :0      Idle
Rudder     :0
Auxiliary  :-100   Off
Arm Test   :Safe Zone
BACK
```

29: TUNING

Stick Scaling—which will increase or decrease the influence each stick has on the 'copter—is the equivalent to *Dual Rate* on a transmitter, but setting it here is better for signal quality. The standard settings are good.

```
Place the aircraft on
a level surface and
press CONTINUE.
The FC will then wait
5 sec to let the
aircraft settle down.
                CONTINUE
```

P&I gains (often called PID settings) should also be checked at this stage, but be aware that you will have to refine them once you've flown the craft. We will look at this in more detail on the following pages.

Finally, you will need to tell the 'copter when it is level via the *ACC Settings* menu option and on-screen instructions.

30: FIT PROPS

Before you fit the props, arm the copter and gently push the throttle up, stopping as soon as they're all turning. See which way each motor spins and, if necessary, swap any two of the cables to the motor until they all spin in the way the step 25 diagram suggests.

Now disarm the 'copter, remove the tape, disconnect the battery, and fit your props. You need two pairs of props (each pair has a clockwise and a counterclockwise prop), and need to take care that you put the correct props on the correct motors, based on the direction of rotation.

MAIDEN FLIGHT

Now it's time to take your 'copter somewhere very safe, big, and empty for your maiden flight. Place it on a flat, level surface, some distance from yourself.

Ease into the air and try to get a feel of the responses to your different movements.

Tweet your results with the hashtag #tamesky1.

POST-BUILD TUNING

Once you've got your 'copter in the air, there's a lot you can do to make it easier to fly and tuning the PID gains is an artform worthy of many hours of discussion (just ask any committed enthusiast).

PID settings are not unique to multicopters and many books and thesis have been written on the subject. They are the elements in the equation of a "control loop," a simple program testing and re-testing a mechanism and constantly tweaking the inputs (controls) to make sure that it maintains performance. For the sake of simplicity, this example is the throttle formula required for one motor; the P and I values are yours to experiment with.

P GAIN

P stands for "Proportional" and it is the extra force used by a 'copter to balance it out when it drifts away from the level. It is proportional because more force is applied if it is further out of level.

When the P gain is set far too low, the 'copter will want to flip over, because not enough correction is applied. If it is just a little low, the 'copter will feel sluggish in flight, and slow to respond to controls. When set to high you'll find the 'copter is jumpy and will wobble in the air because it keeps overshooting with its compensations.

I GAIN

I stands for "Integral," and in this context means "over time." The implementation varies, but in principle it means the additional increase applied to correct an issue depending on how long it's been a factor.

When the I gain is set too low, a 'copter might be more likely to drift, rather than level out after it's been hit by a strong gust of wind. However, this is much less of a problem than when it is too high, so start with a lower value and work upward.

When the I gain is far too high, there will be the same wobble as a high P gain, but it will get worse before it gets better (by which time it'll likely be worse in the opposite direction). Slightly too high and you'll find that there seems to be some push-back to flights.

EXPO

Expo is a transmitter setting that goes a long way toward making a multicopter easier to fly (the setting might be labeled "curves" on your system). Check your manual, and aim to set a curve so that the first 50% of the physical stick movement only has 10% of the influence on the 'copter. This will make flight a lot easier, as the sticks will seem far less sensitive when maneuvering more gently (meaning that landing should be way easier!).

PID FORMULA

$$\chi_2 = \chi_1 + (P \times \alpha) + (I \times \sigma)$$

χ_1 = Throttle before loop α = Degrees out of level
χ_2 = Throttle after loop σ = Time out of level

PROP BALANCING

Balancing props is the process of making sure that one blade of the prop is equally weighted to the other one, so there is no bias. A perfectly balanced prop will fly more smoothly, reducing vibration, which is good news for the motors and can boost flight times.

At the risk of causing a wave of dissent amongst some pilots, the necessity of prop balancing is open to debate—the plastic props used by many fliers are virtually disposable, and are usually supplied in reasonable (if not perfect) balance.

That said, a magnetic prop-balancing tool that allows you to clamp your prop onto a shaft supported only by magnets is not expensive. Perfectionists may want to try balancing the prop's hub as well.

PROP-BALANCING TOOL

A prop-balancing tool allows the prop to be held on a shaft supported by magnets. If the prop stays level, the blades are balanced; if one blade drops, they are not. Once your blades seem OK, try flipping the prop over and seeing if the central hub seems to be weightier on one side or another too.

UNBALANCED PROP

Here the prop is clearly not balanced, with one blade falling quickly to the surface.

BALANCED PROP

The solution is to add a small piece of sticky tape to the top of the lighter blade (never the bottom side).

CHAPTER 06
RESOURCES

DRONES AND THE LAW

Thanks in part to unhelpful media coverage, drones (and in this context the choice of the word is unarguable) have captured the attention of law makers around the world. However, the situation is invariably a little more nuanced than people would have you believe.

AVIATION AUTHORITIES

Until the invention of the aircraft, it was an unchallenged legal principle that people owned all the land beneath their property and all above it, so flying on your own property would never be a problem, even at 12,000ft. That, of course is no longer true—airspace is regulated by the Federal Aviation Administration (FAA) in the US, by the Civil Aviation Authority (CAA) in the UK, the DGCA in France, and similar organizations elsewhere.

For the large part, these agencies are used to dealing with licensing for pilots, from light aircraft up, so inevitably some have struggled to adapt to the arrival of drones sold in large numbers, most notably the American FAA which began with a blanket ban on commercial activity until mandated by Congress to find a workable solution. This has meant that virtually no one's aerial photography business got off the ground until 2014, save for a few companies offered exemptions through Section 333 of the FAA Modernization and Reform Act 2012 (essentially to help the FAA develop the codes in practice).

Other authorities have been quicker to react, such as the British CAA, which has backed a program of courses (taught by external course providers) that certify a pilot as competent. Although these BNUCs are not cheap—expect to pay roughly the same amount as you would for a DJI Inspire One for the BNUC-S (under 20kg category)—they open the door to working commercially, by certifying your competence to fly and providing you with the knowledge required to submit plans for a flight.

In 2015 there has been steady progress toward rational codification of the rules for commercial pilots in the US including a number of consultations in which the flying community—both hobby and commercial—were asked for feedback. It is much better to engage with lawmakers this way, before laws are made and when there is no cost. If you wait until something is in the statute book and find yourself disagreeing with it, you had better be prepared to go up against the government in court or look for a change of the law through lobbyists.

CIVIL LIABILITY

It's important to remember that law has other facets and these agencies are just one of them. Civil liability is another crucial concept—if you cause someone harm, they are likely to be entitled to recover money from you to make good that harm.

In other words, you need to consider your liability, and look to take out public liability insurance. This is usually available at a minimal cost, and will almost certainly be required to participate in organized events. It should go without saying that this kind of insurance will only cover you if you are flying within the terms set by the insurer, and this will inevitably require compliance with other rules affecting hobbyists (proximity to people and buildings, and so on).

PRIVACY

People's right to privacy is protected in a number of ways, both through criminal law (the kind enforced by the police that lands you in jail) and civil actions. A right to privacy is emerging in case law in many places, even where it isn't already in a local bill of rights. The European Convention on Human Rights, Article 8, specifically provides for respect to an individual's private life, while in the US case law is developing similar principles.

LOCAL RESTRICTIONS

A number of cities are filling the vacuum of national legislation by implementing ordinances that ban or restrict flying in certain areas. Of course, in other places there are longstanding restrictions that must be restricted, such as military bases.

ENFORCEMENT

The easiest way to earn yourself a large fine (or worse) is to make life easy for the prosecuting authority. Posting your videos on YouTube and other flight-sharing platforms makes evidence gathering very easy, as many people have already found out.

US DEPARTMENT OF TRANSPORT
Uninspiring office buildings like these, in which the FAA reside, house the real people who help influence the future of drones. Engage with them.

SAFE TO FLY?

Knowing where it is safe to fly isn't just a matter for legislation; you should take note of common sense and weather. That said, there are also some very definite restrictions on where you can fly, and some useful services to help you discover them, so there's no excuse.

FLYING YOUR 'COPTER safely is more important than just avoiding expensive repair bills: it only takes one idiot to fly somewhere dangerous and cause the loss of a life and you can bet that legislation will get a lot more onerous. You may well hypothesize that a 2lb (1kg) 'copter isn't going to bring down a commercial jet, but that's no reason why you should fly it on an airport approach—please keep safe.

WIND

It is never safe to fly if the wind is more than half your 'copters maximum speed, and to be honest I like to keep it below that—windspeed can climb quickly, will not be even, and gusts will be faster than the average speed. A DJI Phantom can travel at about 30mph (48kmph), so 10mph (16kmph) is a sensible maximum windspeed. If you're unsure about windspeed you can measure it at your take off location with an anemometer, and some locations might feature a windsock.

POWER LINES AND MASTS

Power lines are a serious threat; not only are they very easy to hit if you're flying near them, but if they're powerful enough they will create a certain amount of EMI (electromagnetic interference). This can affect your control connection and upset the magnetometers needed for GPS loiter modes. Telecommunications masts (which are even less obvious these days) can also wreak havoc with your flight systems.

WEATHERFLOW
The WeatherFlow anemometer can be plugged into the top of a smartphone to give accurate wind readings.

RESTRICTED AIRSPACE

It is, of course, vital that you don't fly in any of the areas reserved for civil or military aviation, but how do you find them? These days, online sectional charts (aviation maps) are the easiest to use—Skyvector is a great choice if you're in the US (or elsewhere), while

SkyDemonLight.com is good if you're in the UK. Both tools reveal how complicated airspace is, and they're also updated to indicate one-off restrictions such as air shows. Much of the information is extraneous, but if you zoom in tightly and plot a route—remember to indicate your likely maximum altitude—both will indicate if you're passing into an area you shouldn't.

PEOPLE

Never, ever fly over people who aren't aware you're flying there. If you're shooting a movie then it should go without saying the actors and crew are in on it, but flying over a town center, sports field, road, or highway where the people below have no warning is simply unacceptable. Although off-the-shelf 'copters seem infallible, even the most beautifully designed craft is vulnerable to failure. In particular, quadcopters have no back up—one burned out ESC or motor, or a damaged/snapped prop, and it will plummet onto one (or many) unsuspecting victims below.

If anything animals, are even more at risk. I've seen dogs charge up to 'copters as they land and surprise the pilot with their speed. People can easily emerge with dogs unleashed and dogs and others don't always perceive the danger of high-speed spinning props (but you can bet the owner will do and will blame you!).

TREES AND FENCES

The obvious dangers of trees should go without saying, but people don't always consider the risks posed by fences. The simple fact is that if your 'copter passes into an area that's sealed off and something untoward happens to it, then you may well have lost it for good—it's as simple as that.

DJI NO FLY ZONES
DJI aircraft have a built-in database of major airports and other restricted locations. With GPS enabled, they'll refuse to fly within 1.5 miles (2.4km) of a major airport, with progressively higher altitude restrictions from 1.5-5 miles (2.4-8km).

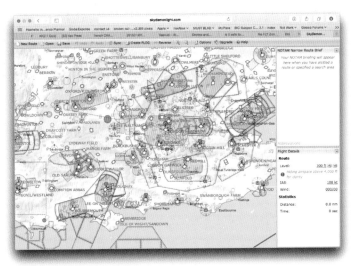

SKYDEMONLITE.COM
Using the online sectional maps at SkyDemon, you can mark where you'll be flying and be alerted to any potential hazards.

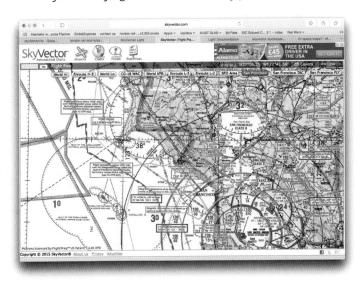

SKYVECTOR
Skyvector.com is a great way to access aeronautical charts.

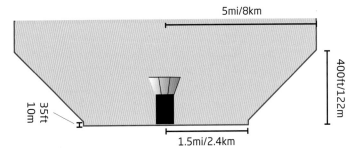

5mi/8km

400ft/122m

35ft 10m

1.5mi/2.4km

COMMUNITIES

Online and IRL (In Real Life) communities make flying a great deal more fun and go a long way to helping you out with the inevitable technical difficulties. Even if you've opted for a ready-to-fly model, there will come a time when it really helps to talk to someone who'd been there before.

COMMUNITIES OF PEOPLE can achieve great things. To give just one example, the team at FirstPersonView. com managed to successfully lobby the UK's Civil Aviation Authority to include an exception in the law (which required pilots line of sight for UAVs) that allowed for goggle-based FPV with a spotter. Without people ready to stand up for their favorite form of flying, it could have been banned by a bureaucrat.

Another is the International Drone Day, an event put together by Sarah and David Oneal, whose video podcast, *That Drone Show* is a great way to keep informed with the latest news in the drone community. Keen to highlight their "drones are good" mantra (against a backdrop of fearmongering media stories), they inspired pilots around the world to organize and publicize meetups, pushing the positives of the drones to the media, and boy did it work. The unstoppable pair is also the driving force behind the crowd-funded documentary, *The Drone Invasion*.

At a more day-to-day level, any group will provide you with the advice and inspiration you need to keep going. When you're building, it's really useful to be able to talk with people about what you're building, so make use of the forums provided on whatever site your flight controller is associated with (see Chapter 2). In my experience, every question I've asked has been answered within a few hours by someone who has encountered the same issue in the past. It should also go without saying that at the TameSky website you'll find links to the ever-growing list of communities (of every kind) too.

RACING LIGHTS
This racing event, organized by fpvleague.co.uk and hosted by RadioC.co.uk, was throughly tricked out, with traffic light start signals. Marshalling here is Dan Reid.

MONITOR

At some events, those not flying can monitor the progress of others via a monitor set to the same frequency as the flying pilot's FPV goggles. This makes for a thoroughly entertaining event, even when the 'copter is out of sight behind trees or obstacles.

TEAM PHOTO

Make sure you group everyone together for a team photo at events and use it to promote the next. (Photo ©Gareth Green)

MEETUPS AND COMPETITIONS

Although you will probably need to take out some form of Public Liability Insurance to fly with others, going to events and flying against the clock will help you sharpen your skills in locations that you might not otherwise be able to.

If there are no such organizations in your area, it's easy to start one using Meetup.com, Facebook's groups feature or any of the other online social networking tools. Decide on a meeting place and a date, and then make sure all the details are on the site. Meetup requires a small fee, but is a very effective way of finding people keen to meet face to face.

COMMUNITY SITES

rc groups A fantastic forum always ready with answers for even niche questions.

flitetest In its own words, "entertaining, educating, and elevating." Covers more than just multicopters, but a great vodcast and community site.

That Drone Show News, reviews, interviews, and competitions, with a lot of coverage from major equipment launches.

TameSky.com This book's online home, with videos of the Chapter 5 project and more.

airgonay.com French flying club.

GLOSSARY

If multicopters have gathered one thing, it's jargon. It seems as if it has been inherited from everywhere—computers, the radio control community, autopilot systems, and more—so let's cut through the tech and the acronyms to see what it really means.

ACCELEROMETER (ACC)

A component that measures acceleration in a given axis (essentially measuring G force).

ARDUINO

A low-cost computer/circuit board with open source (free) software that allows accessories such as servos to be controlled easily. Many flight controllers are built on this system, including ArduCopter.

ARF (also ARTF)

Almost Ready to Fly. Partially pre-built 'copter kits that might be missing a key element (typically a transmitter).

AUTONOMOUS

In the context of drones, a 'copter that is flying itself, typically by following a pre-programed flight plan consisting of multiple waypoints.

BAROMETRIC PRESSURE SENSOR

A sensor that detects altitude by means of the air pressure.

BEC

Battery Eliminator Circuit. This is a circuit that allows one battery to power multiple devices onboard a 'copter. For example, it might allow the battery to power the motors at full voltage, as well as smaller motors on a gimbal, or the power needed for FPV. In each case, the circuit will safely step the power down. Although this increases the draw on the main battery, it requires less weight (and volume) than carrying two batteries, and makes charging easier.

BIND

The process of associating a receiver with a transmitter.

BNF

Bind and Fly. A 'copter that is ready to be bound to a transmitter and flown.

BNUC

The qualification that the CAA in the UK uses to assess a UAV pilot's experience. A cheaper BNUC-S version covers lighter aircraft, but neither is required for hobby use.

BRUSHLESS MOTORS

Almost universally adopted in the multicopter world, these powerful motors are more efficient and longer lasting then their brushed predecessors, and less likely to require gearing.

BUILD

You'll frequently hear pilots referring to their creations by using "build" as a noun, as in: "My build seemed fine until I crashed it into that tree."

CAA

Civil Aviation Authority. The UK body responsible for managing airspace above the country.

CAMERASHIP

A multicopter built with photography as a primary purpose.

CF

Carbon Fiber. Strong, light building material from which many airframes are at least partially constructed.

CG

Center of Gravity. This is an important point on your craft as it should typically be at the center (to avoid undue load on any one motor). You will often be required to place your flight controller on or near the CG, or at least be able to tell the flight controller how far it is from the CG in setup.

CONTROLLER
Often used as shorthand for the pilot's radio-control transmitter (not to be confused with the flight controller).

DRONE
A word with somewhat hijacked meaning (not least by those who refuse to use the word to describe something everyone else would instantly recognize as such). In these pages it means a multicopter/UAV with no military connotations (unless otherwise stated).

ESC
Electronic Speed Controller. A device that sits between the flight controller and the motor to regulate the motor's speed.

EXPO
Expo settings change the servo/motor response from a linear line (30% input on the stick, 30% throttle), to an S curve which is flatter (less sensitive) around the center point.

FAA
Federal Aviation Authority. The US authority that manages airspace and sets rules for the use of all aircraft, including multicopters.

FPV
First Person View. Flying using a camera and monitor or video goggles "though the eyes" of the drone.

GAINS
Usually pilots are referring to their PID settings.

GIMBAL
In the case of drones, this invariably refers to a camera mount that uses motors to stay in the same position relative to the ground, regardless of the 'copter's movements.

GLONASS
Russian equivalent of GPS.

GYRO
Short for *Gyroscope*.

GYROSCOPE
A device that measures orientation. Used by the flight controller for leveling the aircraft.

HOBBY GRADE
Another word for drones that are a step up from toys. Encompasses both kit-builds as well as multicopters, such as the DJI Phantom.

IMU
Inertial Measurement Unit. A combined set of gyroscopes and accelerometers that can determine orientation and stability.

INS
Inertial Navigation System. A method of calculating location based on speed and motion sensors, while GPS is temporarily unavailable.

INTERVALOMETER
A device which can instruct a camera to take a picture every five or ten seconds, for example.

KAPTEIN KUK
A flight control board with a built-in LCD screen. Perfect for modestly priced projects (like the TameSky 1).

LHS
Local Hobby Shop. No more technical than it sounds; a short form that you might see used in discussion forums.

LIPO
Lithium Polymer Battery. Almost every drone uses a LiPo battery, due to its power-to-weight ratio.

LIPO BAG
A fire-resistant bag for keeping batteries in. LiPo batteries can degrade and heat up to dangerous temperatures, so LiPo bags should always be used.

LOS

Line of Sight. In almost every operating environment pilots—especially hobbyists—are required to be able to see their 'copter directly at all times.

mAh

MilliAmp Hours. The capacity of a battery. For example, a 1000 mAh cell (1.0Ah/Amp Hours), would be drained in one hour if a 1 Amp load was placed on it; would last half an hour with a 2 Amp load; 15 minutes with a 4 Amp load, and so on. In real terms, the greater the capacity of the battery, the longer your flight time.

MAV

Micro Air Vehicle. A small UAV.

MAVLINK

A communications protocol used by ArduCopter and ArduPlane autopilots.

MOD

A change from the manufacturer's original design; derived from *modification.*

MULTIWII

A general purpose, open source software project initially developed to allow all kinds of hobbyists to use the sophisticated gyroscopes in the Nintendo Wii controller, which quickly grew to support multirotor aircraft. Now, with those sensors cheap and available, the software is implemented onto flight controller boards that include the sensors.

NAZA

A flight controller produced by DJI for self-build hobbyists. A variant of the NAZA system sits inside the company's popular Phantom 1 and 2 series.

NMEA

The initials for the US *National Marine Electronics Association.* Crops up in the term "NMEA Sentences," which refers to an ASCII string from a GPS module.

OCTOCOPTER

A multicopter with eight props.

OPTICAL FLOW SENSOR

A sensor that uses a downward facing camera to identify visible texture/features and use these to measure the speed the drone is traveling over the ground. DJI has made much play of these features, but other craft were using them long before (notably, Parrot's AR.Drone 2).

OSD

On Screen Display. A way to integrate telemetry data into the video link from a drone, so it can be seen in FPV goggles or on a monitor.

PAYLOAD

The amount of weight your vehicle may be able to lift, aside from itself and its batteries.

PIC

Pilot In Control (as opposed to Computer In Control, which would mean autopilot). This is more commonly pro jargon.

PID

Proportional, Integral, and Derivative "gain" settings. These are settings that affect the way the 'copter responds to control input and external factors (wind). P gain is the amount of correction made to bring your 'copter level; I gain the amount of time before the flight controller will try harder to make the correction; D gain will work to harmonize the effects of the P and I settings as you are approaching level, to avoid over-shooting. The downside is that a high D gain can create a long delay between stick input and action because of the dampening effect.

PITCH

Describes the front/back tilt of the 'copter, which effectively amounts to the forward/backward control on a multicopter.

RC

Radio Controlled. A generic term for all things controlled using a radio controller. The "RC community" encompasses model airplanes, boats, cars, and more.

READY TO FLY (RTF)

A multicopter that is supplied ready to fly, out of the box. There might still be a few settings to adjust, but other than that, you'll be good to go.

RETURN TO HOME (RTH)

A flight mode that takes the aircraft back to the point it took off from, where it will land. This is often set up as a failsafe should the 'copter lose radio connection. Note that it is important to set RTH with an altitude that will clear any nearby obstacles before landing.

RETURN TO LAND (RTL)

See *Return to Home*.

ROLL

The rotation around a point. In an airplane it would be roll around the fuselage (one wing tips up, the other down); in a multicopter it is leftward or rightward movement without moving turning the front (as opposed to yaw/rudder).

Rx

Short for "receiver" or "receive."

TELEMETRY

Data received from a remote system, typically including information on speed, altitude, battery status, and so on. Many Rx/Tx systems include built-in telemetry systems. Telemetry is taken for granted by pilots of systems such as the Phantom 3, but it is by no means necessary for flight.

THROTTLE

Used to increase or decrease the rotation of the props, the throttle is effectively an up/down control.

TOY GRADE

Hobby retailers like to distinguish between "hobby grade" (good/expensive) and "toy grade" (poor/low-cost) multicopters, but much can be learned from the art of keeping a "toy grade" microcopter in the air. However,

don't expect to be able to replace parts or take quality video with models in this category.

Tx

Short for "transmitter" or "transmit."

UAS

Unmanned Aerial System. A term derived from a military adaptation of UAV, which is meant to better reflect that the craft is just part of a chain, with infrastructure also required on the ground.

UAV

Unmanned Aerial Vehicle. Any aircraft that is flown remotely (though occasionally confused with Unmanned Autonomous Vehicle, which can refer to anything, airborne or not, and isn't usually a civilian term).

ULTRASONIC SENSOR

Used to determine distance from the ground by bouncing soundwaves off it. Only works reliably at distances of approximately 6½–10ft (2–3m).

VLOS

Visual Light of Sight. See *LOS*.

WAYPOINT

A specific location in space, defined by a set of coordinates. If you're setting an autonomous flight, you'll typically send the drone to a series of waypoints.

WOT

Wide Open Throttle. 100% throttle, as in: "My sweet 'copter will shoot up 10 meters a second WOT."

YAW

Describes the rotation around a central point. On an airplane, the turning effect is provided by the rudder.

INDEX

Yours to photocopy then cut out when building the TameSky1. Alternatively, you can download and print your own "wooden wonder" at www.tamesky.com

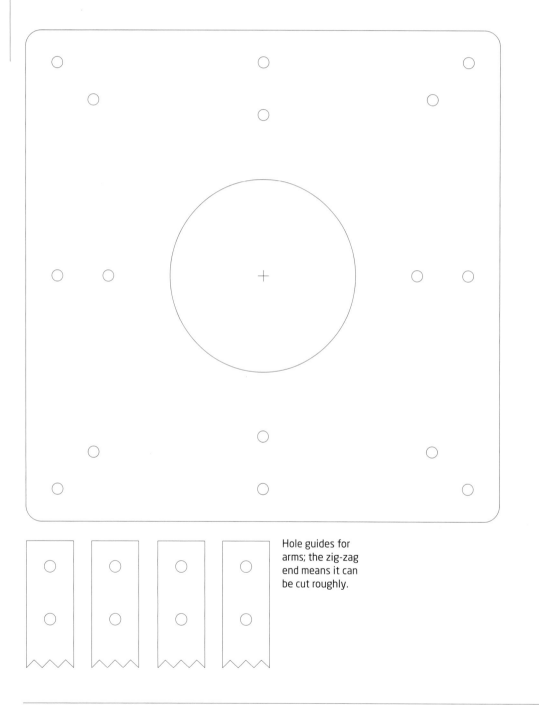

Leg 1

Leg 2

Hole guides for arms; the zig-zag end means it can be cut roughly.

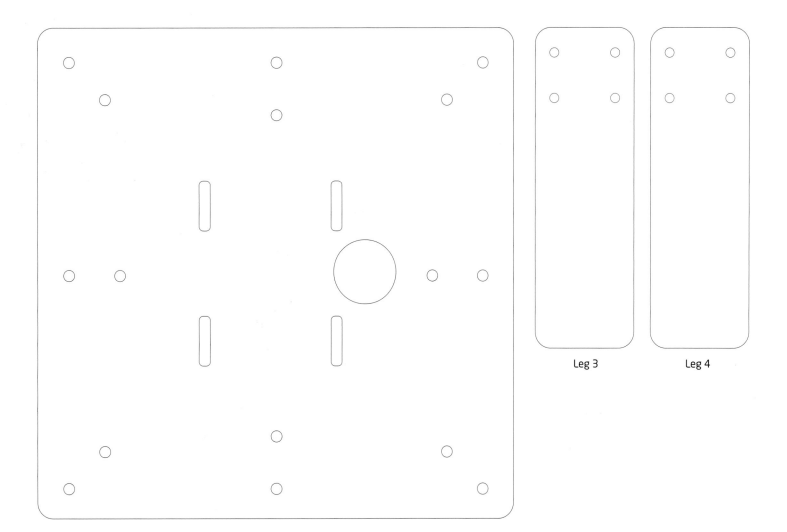

Leg 3   Leg 4

ACKNOWLEDGMENTS

06

RESOURCES

I wouldn't have been able to write this book without the assistance of a great many people, whether that is members of the user groups who shared their photos with me or, well, anyone who has had to put up with me in the past few months. Cap duly doffed to Chris Gatcum for some super-fast editing, Frank Gallaugher at Ilex for ensuring things went smoothly (despite my delayed manuscript), Roly Allen for nudging me in the direction of this book in the first place, and Julie Weir & Jon Allan for the speedy design work (and especially to Julie for her much-appreciated patience). I also mustn't forget Rachel Silverlight for her lightning edits and the international rights team at Octopus.

Thanks must also go to Fabiano Calcioli and the team at RadioC, Thomas Greer of LondonAerospace for answering a lot of questions (and letting me play with the traffic lights), Gareth Greener who has shared some great photos, David Oneal of *That Drone Show* for a few pointers, and to the team at Brighton-based FirstPersonView.co.uk. Indeed everyone I met with or contacted during my research was always incredibly helpful, and I only wish there was space to thank everyone as profusely as they deserve.

This book wouldn't have been what it is at all, though, without Jack Nash (BrightonTiliFly) who helped me see that the 5th chapter would work so much better with a build, rather than a random set of skills, and who—along with Niko Arasaki & I—did the lion's share of the work. That includes design, prototype build, maiden flight, and of course testing the first prototype to destruction!.

Finally, of course, thanks to my family, especially my father for a thorough read-through, and for the wonderful woman who got me hooked in the first place. Vasiliki. Σε αγαπώ.